VOLCANOES AND EARTHQUAKES

Britannica Illustrated Science Library

Encyclopædia Britannica, Inc.
Chicago ■ London ■ New Delhi ■ Paris ■ Seoul ■ Sydney ■ Taipei ■ Tokyo

Britannica Illustrated
Science Library

© **2009 Editorial Sol 90**
All rights reserved.

Idea and Concept of This Work: Editorial Sol 90

Project Management: Fabián Cassan

Photo Credits: Corbis, ESA, Getty Images, Graphic News, NASA, National Geographic, Science Photo Library

Illustrators: Guido Arroyo, Pablo Aschei, Gustavo J. Caironi, Hernán Cañellas, Leonardo César, José Luis Corsetti, Vanina Farías, Joana Garrido, Celina Hilbert, Isidro López, Diego Martín, Jorge Martínez, Marco Menco, Ala de Mosca, Diego Mourelos, Eduardo Pérez, Javier Pérez, Ariel Piroyansky, Ariel Roldán, Marcel Socías, Néstor Taylor, Trebol Animation, Juan Venegas, Coralia Vignau, 3DN, 3DOM studio

Composition and Pre-press Services: Editorial Sol 90

Translation Services and Index: Publication Services, Inc.

Britannica Illustrated Science Library Staff

Editorial
Michael Levy, *Executive Editor, Core Editorial*
John Rafferty, *Associate Editor, Earth Sciences*
William L. Hosch, *Associate Editor, Mathematics and Computers*
Kara Rogers, *Associate Editor, Life Sciences*
Rob Curley, *Senior Editor, Science and Technology*
David Hayes, *Special Projects Editor*

Art and Composition
Steven N. Kapusta, *Director*
Carol A. Gaines, *Composition Supervisor*
Christine McCabe, *Senior Illustrator*

Media Acquisition
Kathy Nakamura, *Manager*

Copy Department
Sylvia Wallace, *Director*
Julian Ronning, *Supervisor*

Information Management and Retrieval
Sheila Vasich, *Information Architect*

Production Control
Marilyn L. Barton

Manufacturing
Kim Gerber, *Director*

Encyclopædia Britannica, Inc.

Jacob E. Safra, *Chairman of the Board*

Jorge Aguilar-Cauz, *President*

Michael Ross, *Senior Vice President, Corporate Development*

Dale H. Hoiberg, *Senior Vice President and Editor*

Marsha Mackenzie, *Director of Production*

International Standard Book Number (set):
978-1-59339-845-3
International Standard Book Number (volume):
978-1-59339-400-4
Britannica Illustrated Science Library:
Volcanoes and Earthquakes 2009

Printed in China

www.britannica.com

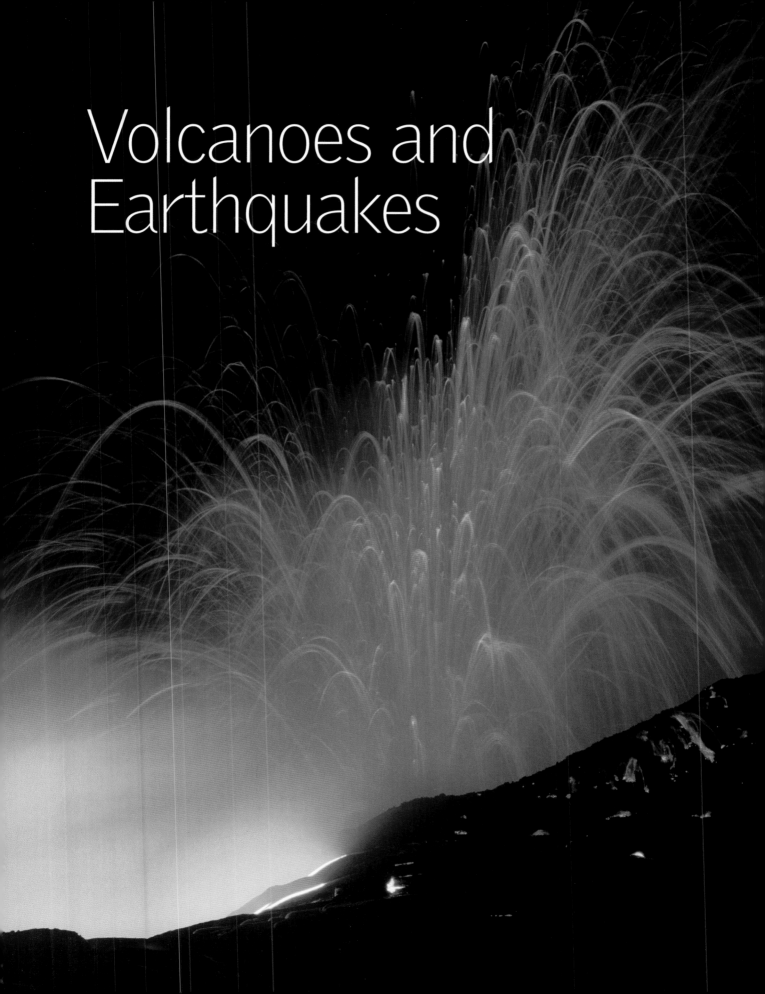

Volcanoes and Earthquakes

Contents

Kashmir, 2005
Farmer Farid Hussain, 50, grasps the hand of his wife, Akthar Fatma, after the earthquake that rocked the Himalayas on the Indian subcontinent. Eighty thousand people were killed, and thousands of families were left homeless.

The Power of Nature

S ome photos speak for themselves. Some gestures communicate more than words ever could, like these clasped hands, which seek comfort in the face of fear of the unknown. The picture was taken Oct. 8, 2005, when aftershocks were still being felt from the strongest earthquake ever to strike Kashmir, in northern India. Those clasped hands symbolize terror and panic; they speak of fragility and helplessness, of endurance in the face of chaos. Unlike storms and volcanic eruptions, earthquakes are unpredictable, unleashed within seconds, and without warning. They spread destruction and death, forcing millions to flee from their homes. The day after the catastrophe revealed a terrifying scene: debris everywhere, a number of people injured and dead, others wandering desperately, children crying, and over three million survivors seeking help after losing everything. Throughout history Earth has been shaken by earthquakes of greater or lesser violence. These earthquakes have caused great harm. One of the most famous is the earthquake that rocked San Francisco in 1906. Registering 8.3 on the Richter scale, the temblor left nearly three thousand dead and was felt as far away as Oregon to the north, and Los Angeles in southern California.

The purpose of this book is to help you better understand the causes of fractures and the magnitude and violence of the forces deep within the earth. The full-color, illustrated book you hold in your hands contains shocking scenes of cities convulsed by earthquakes and volcanoes, natural phenomena that, in mere seconds, unleash rivers of fire, destroy buildings, highways and bridges, and gas and water lines and leave entire cities without electricity or phone service. If fires cannot be put out quickly, the results are even more devastating. Earthquakes near coastlands can cause tsunamis, waves that spread across the ocean with the speed of an airplane. A tsunami that reaches a coast can be more destructive than the earthquake itself. On Dec. 26, 2004, the world witnessed one of the most impressive natural disasters ever. An undersea quake with a magnitude of 9 on the Richter scale shook the eastern Indian Ocean, causing tsunamis that reached the coastal areas of eight Asian nations, causing about 230,000 deaths. The earthquake was the fifth strongest since the invention of the seismograph. Satellite images show the region before and after the catastrophe.

Throughout history, nearly all ancient peoples and large societies have thought of volcanoes as dwelling places of gods or other supernatural beings to explain the mountains' fury. Hawaiian mythology, for instance, spoke of Pele, the goddess of volcanoes, who threw out fire to cleanse the earth and fertilize the soil. She was believed to be a creative force. Nowadays, specialists try to find out when a volcano might start to erupt, because within hours after an eruption begins, lava flows can change a lush landscape into a barren wilderness. Not only does hot lava destroy everything in its path, but gas and ash expelled in the explosion also replace oxygen in the air, poisoning people, animals, and plants. Amazingly, life reemerges once again from such scenes of destruction. After a time, lava and ash break down, making the soil unusually fertile. For this reason many farmers and others continue to live near these "smoking mountains," in spite of the latent danger. Perhaps by living so close to the danger zone, they have learned that no one can control the forces of nature, and the only thing left to do is to simply live.●

Continuous Movement

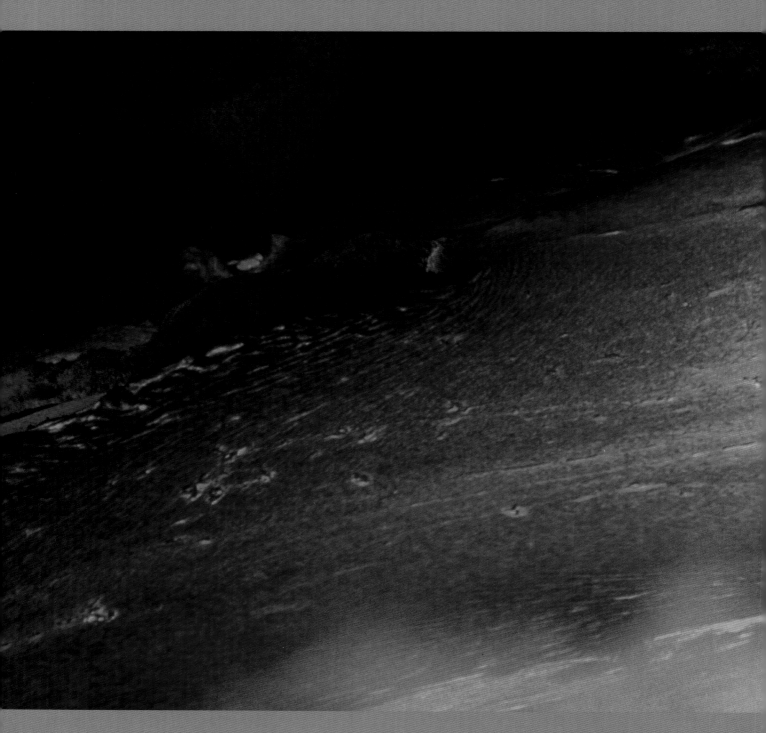

In the volatile landscape of Volcano National Park in Hawaii, the beginning and end of life seem to go hand in hand. Outpourings of lava often reach the sea. When the molten rock enters the water, the lava quickly cools and hardens into rock that becomes part of the coastline. By this process, volcanic islands grow constantly, and

PAHOEHOE LAVA
A type of Hawaiian lava
that flows down the slopes
of Mt. Kilauea to the sea.

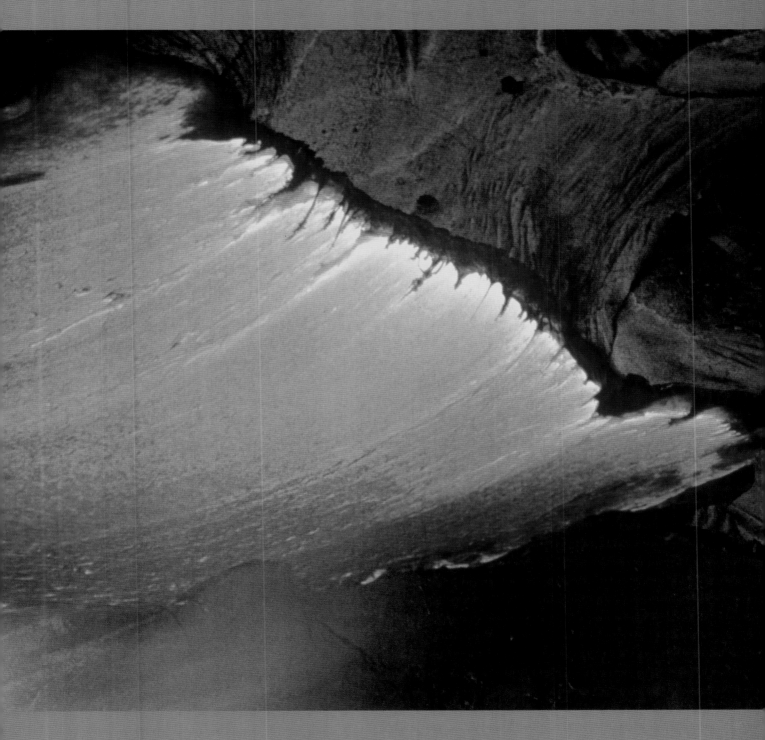

nothing stays the same from one moment to another. One day rivers of lava blaze down the volcano's slopes, and the next day there are new, silver-colored rocks. The ongoing investigation of lava samples under the microscope helps volcanologists discover the rock's mineral composition and offers clues about how the volcano may behave. ●

Scorching Flow

ost of the Earth's interior is in a liquid and incandescent state at extremely high temperatures. This vast mass of molten rock contains dissolved crystals and water vapor, among other gases, and it is known as magma. When part of the magma rises toward the Earth's surface, mainly through volcanic activity, it is called lava. As soon as it reaches the surface of the Earth or the ocean floor, the lava starts to cool and solidify into different types of rock, according to its original chemical composition. This is the basic process that formed the surface of our planet, and it is the reason the Earth's surface is in constant flux. Scientists study lava to understand our planet better. ●

Streams of Fire

Lava is at the heart of every volcanic eruption. The characteristics of lava vary, depending on the gases it contains and its chemical composition. Lava from an eruption is loaded with water vapor and gases such as carbon dioxide, hydrogen, carbon monoxide, and sulfur dioxide. As these gases are expelled, they burst into the atmosphere, where they create a turbulent cloud that sometimes discharges heavy rains. Fragments of lava expelled and scattered by the volcano are classified as bombs, cinders, and ash. Some large fragments fall back into the crater. The speed at which lava travels depends to a great extent on the steepness of the sides of the volcano. Some lava flows can reach 90 miles (145 km) in length and attain speeds of up to 30 miles per hour (50 km/hr).

INTENSE HEAT
Lava can reach temperatures above 2,200° F (1,200° C). The hotter the lava, the more fluid it is. When lava is released in great quantities, it forms rivers of fire. The lava's advance is slowed down as the lava cools and hardens.

Mineral Composition

Lava contains a high level of silicates, light rocky minerals that make up 95 percent of the Earth's crust. The second most abundant substance in lava is water vapor. Silicates determine lava's viscosity, that is, its capacity to flow. Variations in viscosity have resulted in one of the most commonly used classification systems of lava: basaltic, andesitic, and rhyolitic, in order from least to greatest silicate content. Basaltic lava forms long rivers, such as those that occur in typical Hawaiian volcanic eruptions, whereas rhyolitic lava tends to erupt explosively because of its poor fluidity. Andesitic lava, named after the Andes mountains, where it is commonly found, is an intermediate type of lava of medium viscosity.

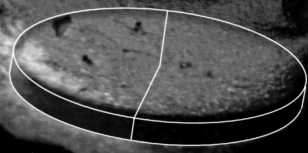

52%
Silicates

48%
Other Content

Rock Cycle

Once it cools, lava forms igneous rock. This rock, subjected to weathering and natural processes such as metamorphism and sedimentation, will form other types of rocks that, when they sink back into the Earth's interior, again become molten rock. This process takes millions of years and is known as the rock cycle.

SEDIMENTARY ROCK
Rock formed by eroded and compacted materials.

METAMORPHIC ROCKS
Their original structure is changed by heat and pressure.

TURNS BACK INTO LAVA

TURNS BACK INTO LAVA

2. **IGNEOUS ROCK**
Rock formed when lava solidifies. Basalt and granite are good examples of igneous rocks.

1. **LAVA**
The state in which magma flows to the Earth's outer crust, either reaching the surface or getting trapped within the crust.

SOLID LAVA
Lava solidifies at temperatures below 1,700° F (900° C). The most viscous type of lava forms a rough landscape, littered with sharp rocks; more fluid lava, however, tends to form flatter and smoother rocks.

1,800° F
(1,000° C)

is the average temperature of liquid lava.

TYPES OF LAVA
Basaltic lava is found mainly in islands and in mid-ocean ridges; it is so fluid that it tends to spread as it flows. Andesitic lava forms layers that can be up to 130 feet (40 m) thick and that flow very slowly, whereas rhyolitic lava is so viscous that it forms solid fragments before reaching the surface.

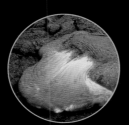

Andesitic Lava

Silicates	63%
Other Content	37%

Rhyolitic Lava

Silicates	68%
Other Content	32%

The Long History of the Earth

The nebular hypothesis developed by astronomers suggests that the Earth was formed in the same way and at the same time as the rest of the planets and the Sun. It all began with an immense cloud of helium and hydrogen and a small portion of heavier materials 4.6 billion years ago. Earth emerged from one of these "small" revolving clouds, where the particles constantly collided with one another, producing very high temperatures. Later, a series of processes took place that gave the planet its present shape. ●

From Chaos to Today's Earth

Earth was formed 4.6 billion years ago. In the beginning it was a body of incandescent rock in the solar system. The first clear signs of life appeared in the oceans 3.6 billion years ago, and since then life has expanded and diversified. The changes have been unceasing, and, according to experts, there will be many more changes in the future.

4.5
BILLION YEARS AGO

COOLING

The first crust formed as it was exposed to space and cooled. Earth's layers became differentiated by their density.

4.6 BILLION YEARS AGO

FORMATION

The accumulation of matter into solid bodies, a process called accretion, ended, and the Earth stopped increasing in volume.

60 MILLION YEARS AGO

FOLDING IN THE TERTIARY PERIOD

The folding began that would produce the highest mountains that we now have (the Alps, the Andes, and the Himalayas) and that continues to generate earthquakes even today.

540 MILLION YEARS AGO

PALEOZOIC ERA

FRAGMENTATION

The great landmass formed that would later fragment to provide the origin of the continents we have today. The oceans reached their greatest rate of expansion.

1.0 BILLION YEARS AGO

SUPERCONTINENTS

Rodinia, the first supercontinent, formed, but it completely disappeared about 650 million years ago.

4
BILLION YEARS AGO
METEORITE COLLISION

Meteorite collisions, at a rate 150 times as great as that of today, evaporated the primitive ocean and resulted in the rise of all known forms of life.

3.8
BILLION YEARS AGO
ARCHEAN EON
STABILIZATION

The processes that formed the atmosphere, the oceans, and protolife intensified. At the same time, the crust stabilized, and the first plates of Earth's crust appeared. Because of their weight, they sank into Earth's mantle, making way for new plates, a process that continues today.

When the first crust cooled, intense volcanic activity freed gases from the interior of the planet, and those gases formed the atmosphere and the oceans.

THE AGE OF THE SUPER VOLCANOES

Indications of komatite, a type of igneous rock that no longer exists.

The oldest rocks appeared.

2.2
BILLION YEARS AGO
WARMING

Earth warmed again, and the glaciers retreated, giving way to the oceans, in which new organisms would be born. The ozone layer began to form.

2.3
BILLION YEARS AGO
"SNOWBALL" EARTH

Hypothesis of a first, great glaciation.

1.8
BILLION YEARS AGO
PROTEROZOIC EON
CONTINENTS

The first continents, made of light rocks, appeared. In Laurentia (now North America) and in the Baltic, there are large rocky areas that date back to that time.

Stacked Layers

E very 110 feet (33 m) below the Earth's surface, the temperature increases by 1.8 degrees Fahrenheit (1 degree Celsius). To reach the Earth's center—which, in spite of temperatures above 12,000° F (6,700° C), is assumed to be solid because of the enormous pressure exerted on it—a person would have to burrow through four well-defined layers. The gases that cover the Earth's surface are also divided into layers with different compositions. Forces act on the Earth's crust from above and below to sculpt and permanently alter it. ●

Earth's crust

Earth's crust is its solid outer layer, with a thickness of 3 to 9 miles (4 to 15 km) under the oceans and up to 44 miles (70 km) under mountain ranges. Volcanoes on land and volcanic activity in the mid-ocean ridges generate new rock, which becomes part of the crust. The rocks at the bottom of the crust tend to melt back into the rocky mantle.

KEY ● Sedimentary Rock ● Igneous Rock ● Metamorphic Rock

THE CONTINENTAL SHELF
In the area where the oceanic crust comes in contact with a continent, igneous rock is transformed into metamorphic rock by heat and pressure.

THE MID-OCEAN RIDGES
The ocean floor is regenerated with new basaltic rock formed by magma that solidifies in the rifts that run along mid-ocean ridges.

OCEANIC ISLANDS
Some sedimentary rocks are added to the predominantly igneous rock composition.

THE SOLID EXTERIOR
The crust is made up of igneous, sedimentary, and metamorphic rock, of various typical compositions, according to the terrain.

MOUNTAIN RANGES
Made up of the three types of rock in about equal parts.

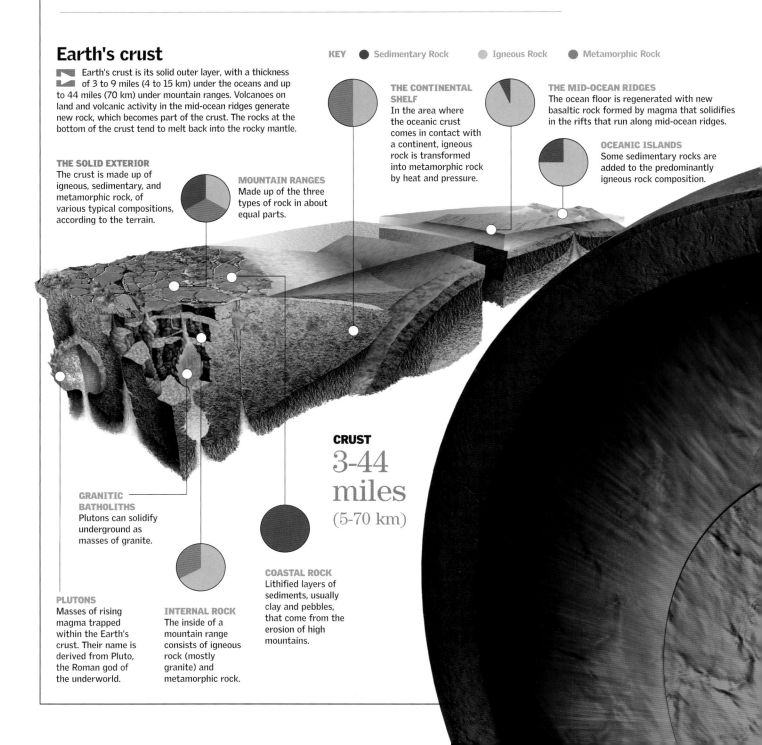

CRUST
3-44 miles
(5-70 km)

GRANITIC BATHOLITHS
Plutons can solidify underground as masses of granite.

PLUTONS
Masses of rising magma trapped within the Earth's crust. Their name is derived from Pluto, the Roman god of the underworld.

INTERNAL ROCK
The inside of a mountain range consists of igneous rock (mostly granite) and metamorphic rock.

COASTAL ROCK
Lithified layers of sediments, usually clay and pebbles, that come from the erosion of high mountains.

The Gaseous Envelope

The air and most of the weather events that affect our lives occur only in the lower layer of the Earth's atmosphere. This relatively thin layer, called the troposphere, is up to 10 miles (16 km) thick at the equator but only 4 miles (7 km) thick at the poles. Each layer of the atmosphere has a distinct composition.

Less than
6 miles
(10 km)
TROPOSPHERE
Contains 75 percent of the gas and almost all of the water vapor in the atmosphere.

Less than
31 miles
(50 km)
STRATOSPHERE
Very dry; water vapor freezes and falls out of this layer, which contains the ozone layer.

Less than
62 miles
(100 km)
MESOSPHERE
The temperature is -130° F (-90° C), but it increases gradually above this layer.

Less than
310 miles
(500 km)
THERMOSPHERE
Very low density. Below 155 miles (250 km) it is made up mostly of nitrogen; above that level it is mostly oxygen.

Greater than
310 miles
(500 km)
EXOSPHERE
No fixed outer limit. It contains lighter gases such as hydrogen and helium, mostly ionized.

UPPER MANTLE
370 miles
(600 km)

LOWER MANTLE
1,430 miles
(2,300 km)
Composition similar to that of the crust, but in a liquid state and under great pressure, between 1,830° and 8,130° F (1,000° and 4,500° C).

OUTER CORE
1,410 miles
(2,270 km)
Composed mainly of molten iron and nickel among other metals at temperatures above 8,500° F (4,700° C).

LITHOSPHERE
Includes the solid outer part of the upper mantle, as well as the crust.

93 miles
(150 km)

ASTHENOSPHERE
Underneath is the asthenosphere, made up of partially molten rock.

280 miles
(450 km)

INNER CORE
756 miles
(1,216 km)
The inner core behaves as a solid because it is under enormous pressure.

The Journey of the Plates

When geophysicist Alfred Wegener suggested in 1910 that the continents were moving, the idea seemed fantastic. There was no way to explain the idea. Only a half-century later, plate tectonic theory was able to offer an explanation of the phenomenon. Volcanic activity on the ocean floor, convection currents, and the melting of rock in the mantle power the continental drift that is still molding the planet's surface today. ●

Continental Drift

The first ideas on continental drift proposed that the continents floated on the ocean. That idea proved inaccurate. The seven tectonic plates contain portions of ocean beds and continents. They drift atop the molten mantle like sections of a giant shell. Depending on the direction in which they move, their boundaries can converge (when they tend to come together), diverge (when they tend to separate), or slide horizontally past each other (along a transform fault).

The Hidden Motor

Convection currents of molten rock propel the crust. Rising magma forms new sections of crust at divergent boundaries. At convergent boundaries, the crust melts into the mantle. Thus, the tectonic plates act like a conveyor belt on which the continents travel.

...180 MILLION YEARS AGO

The North American Plate has separated, as has the Antarctic Plate. The supercontinent Gondwana (South America and Africa) has started to divide and form the South Atlantic. India is separating from Africa.

250 MILLION YEARS AGO

The landmass today's continents come from was a single block (Pangea) surrounded by the ocean.

LAURASIA

GONDWANA

ANTARCTICA

PANGEA

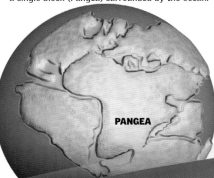

2 inches (5 cm)

Typical distance the plates travel in a year.

CONVERGENT BOUNDARY
When two plates collide, one sinks below the other, forming a subduction zone. This causes folding in the crust and volcanic activity.

Tongan Trench

Eastern Pacific Ridge

Nazca Plate

Peru-C Trenc

Indo-Australian Plate

CONVECTION CURRENTS
The hottest molten rock rises; once it rises, it cools and sinks again. This process causes continuous currents in the mantle.

OUTWARD MOVEMENT
The action of the magma causes the tectonic plate to move toward a subduction zone at its far end.

...100 MILLION YEARS AGO

The Atlantic Ocean has formed. India is headed toward Asia, and when the two masses collide, the Himalayas will rise. Australia is separating from Antarctica.

... 60 MILLION YEARS AGO

The continents are near their current location. India is beginning to collide with Asia. The Mediterranean is opening, and the folding is already taking place that will give rise to the highest mountain ranges of today.

250 MILLION YEARS

The number of years it will take for the continents to drift together again.

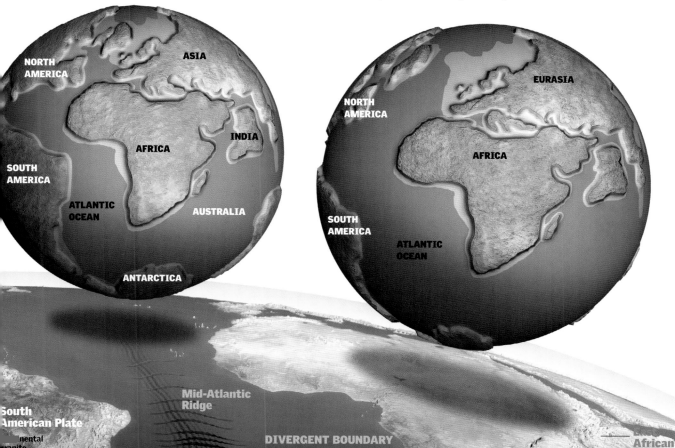

NORTH AMERICA

ASIA

AFRICA

INDIA

SOUTH AMERICA

ATLANTIC OCEAN

AUSTRALIA

ANTARCTICA

EURASIA

NORTH AMERICA

AFRICA

SOUTH AMERICA

ATLANTIC OCEAN

South American Plate

Continental granite

Mid-Atlantic Ridge

DIVERGENT BOUNDARY
When two plates separate, a rift is formed between them. Magma exerts great pressure, and it renews the ocean floor as it solidifies. The Atlantic Ocean was formed in this way.

African Plate

Great African Rift Valley

Somalian Subplate

Subduction zone

Continental crust

WIDENING
At divergent plate boundaries the magma rises, forming new oceanic crust. Folding occurs where plates converge.

Cracks in the Ocean Floor

The concept that the ocean floor is spreading was studied for many years: new crust constantly forms at the bottom of the ocean. The ocean floor has deep trenches, plains, and mountain ranges. The mountain ranges are higher than those found on the continents but with different characteristics. The spines of these great mountain ranges, called mid-ocean ridges, exhibit incredible volcanic activity in rift zones. The rift zones are fissures in relatively narrow regions of the crust, along which the crust splits and spreads. One hundred eighty million years ago, the paleocontinent Gondwana broke apart, forming a rift from which the Atlantic Ocean grew, and is still growing. ●

The Crust Under the Oceans

The constant generation of new ocean crust along rift zones powers a seemingly endless process that generates new lithosphere that is carried from the crest of the ridges, as if on a conveyor belt. Because of this, scientists have calculated that in about 250 million years, the continents will again join and form a new Pangea as they are pushed by the continually expanding ocean floor. Ocean plates are in contact with land plates at the active boundaries of subduction zones or at passive continental boundaries (continental shelves and slopes). Undersea subduction zones, called ocean trenches, also occur between oceanic plates: these are the deepest places on the planet.

HEIGHTS AND DEPTHS

Deep-ocean basins cover 30 percent of the Earth's surface. The depth of the ocean trenches is greater than the height of the greatest mountain ranges, as shown in the graphic below at left.

29,035 feet (8,850 m)
Highest point
(Mount Everest)

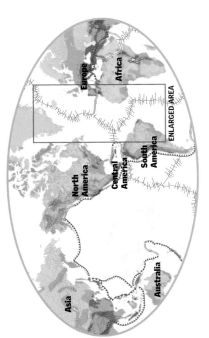

Inside and Outside the Ridge

The abyssal (deep-ocean) plains of the Atlantic are the flattest surfaces on Earth; for thousands of miles, the elevation varies by only about 10 feet (3 m). The plains are made mostly of sediment. Variations in the ocean's depth are mainly the result of volcanic activity, not just within the mid-Atlantic Ridge but elsewhere as well.

1 ATOLLS
Also called coral reefs, atolls are formations of coral deposited around a volcanic cone in warm seas. They form ring-shaped islands.

SOUTH AMERICA

2,900 feet (870 m)
Average land elevation

0 feet (0 m)
Sea level

7,900 feet (2,400 m)
Earth's average elevation

12,240 feet (3,730 m)
Average depth

Normal magnetism

Reversed magnetism

MAGNETISM

Magnetic Reversals

The Earth's magnetic field changes direction periodically. The magnetic north pole changes places with the magnetic south pole. Rock that solidified during a period of magnetic polarity reversal was magnetized with a polarity opposite that of newly forming rocks. Rocks whose magnetism corresponds to the present direction of the Earth's magnetic field are said to have normal polarity, whereas those with the opposite magnetic polarity are said to have reversed polarity.

2 **OCEAN MOUNTS**
Isolated volcanic cones. Some rise above the ocean's surface to become islands, such as the Azores.

How the Mid-Ocean Ridge Was Formed

A spongy layer of rock several dozen miles wide rises above the rift. As the layer fractures and moves away from the fissure, it solidifies into angled blocks that are parallel to the fissure and separated by dikes. Thus the ocean widens as the ridge spreads. The magma exists in a fluid form 2 miles (3.5 km) below the crest of the ridge.

Volcanic smoke

Pillow lava

Fumarole

Dikes within host rock

Oceanic lithosphere

Rising magma

Asthenosphere

Greatest depth
(Mariana Trench)
**About 36,000 feet
(11,000 m)**

Folding in the Earth's Crust

T he movement of tectonic plates causes distortions and breaks in the Earth's crust, especially in convergent plate boundaries. Over millions of years, these distortions produce larger features called folds, which become mountain ranges. Certain characteristic types of terrain give clues about the great folding processes in Earth's geological history. ●

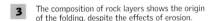

Distortions of the Crust

The crust is composed of layers of solid rock. Tectonic forces, resulting from the differences in speed and direction between plates, make these layers stretch elastically, flow, or break. Mountains are formed in processes requiring millions of years. Then external forces, such as erosion from wind, ice, and water, come into play. If slippage releases rock from the pressure that is deforming it elastically, the rock tends to return to its former state and can cause earthquakes.

1 A portion of the crust subjected to a sustained horizontal tectonic force is met by resistance, and the rock layers become deformed.

2 The outer rock layers, which are often more rigid, fracture and form a fault. If one rock boundary slips underneath another, a thrust fault is formed.

3 The composition of rock layers shows the origin of the folding, despite the effects of erosion.

The Three Greatest Folding Events

The Earth's geological history has included three major mountain-building processes, called "orogenies." The mountains created during the first two orogenies (the Caledonian and the Hercynian) are much lower today because they have undergone millions of years of erosion.

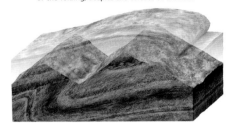

MATERIALS Mudstone, slate, and sandstone, in lithified layers.

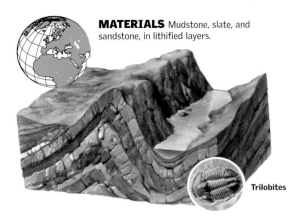

Trilobites

MATERIALS Mostly granite, slate, amphibolite, gneiss, quartzite, and schist.

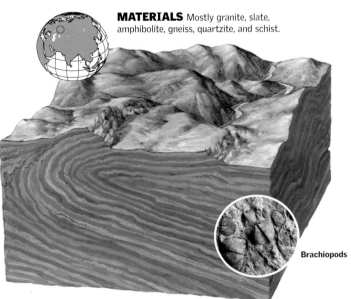

Brachiopods

300 Million Years

HERCYNIAN OROGENY
Took place between the late Devonic and the early Permian periods. It was more important than the Caledonian Orogeny. It shaped central and western Europe and produced large veins of iron ore and coal. This orogeny gave rise to the Ural Mountains, the Appalachian range in North America, part of the Andes, and Tasmania.

430 Million Years

CALEDONIAN OROGENY
Formed the Caledonian range. Remnants can be seen in Scotland, the Scandinavian Peninsula, and Canada (which all collided at that time).

Formation of the Himalayas

The highest mountains on Earth were formed following the collision of India and Eurasia. The Indian Plate is sliding horizontally underneath the Asiatic Plate. A sedimentary block trapped between the plates is cutting the upper part of the Asiatic Plate into segments that are piling on top of each other. This folding process gave rise to the Himalayan range, which includes the highest mountain on the planet, Mount Everest (29,035 feet [8,850 m]). This deeply fractured section of the old plate is called an accretion prism. At that time, the Asian landmass bent, and the plate doubled in thickness, forming the Tibetan plateau.

SOUTHEAST ASIA

India today

10 MILLION YEARS AGO

20 MILLION YEARS AGO

30 MILLION YEARS AGO

Amonites

60 MILLION YEARS

ALPINE OROGENY
Began in the Cenozoic Era and continues today. This orogeny raised the entire system of mountain ranges that includes the Pyrenees, the Alps, the Caucasus, and even the Himalayas. It also gave the American Rockies and the Andes Mountains their current shape.

MATERIALS
High proportions of sediment in Nepal, batholiths in the Asiatic Plate, and intrusions of new granite: iron, tin, and tungsten.

A COLLISION OF CONTINENTS

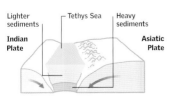

Lighter sediments — Tethys Sea — Heavy sediments

Indian Plate   **Asiatic Plate**

Heavy sediments — Tethys Sea — Tibet

Heavy sediments — Tibet

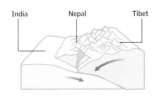

India — Nepal — Tibet

60 MILLION YEARS AGO
The Tethys Sea gives way as the plates approach. Layers of sediment begin to rise.

40 MILLION YEARS AGO
As the two plates approach each other, a subduction zone begins to form.

20 MILLION YEARS AGO
The Tibetan plateau is pushed up by pressure from settling layers of sediment.

THE HIMALAYAS TODAY
The movement of the plates continues to fold the crust, and the land of Nepal is slowly disappearing.

Folds

The force that forms the mountains also molds the rocks within them. As the result of millions of years of pressure, the layers of crust fold into strange shapes. The Caledonian Orogeny, which began 450 million years ago, created a long mountain range that joined the Appalachian mountains of the United States to the Scandinavian peninsula. All of northern England was lifted up during this process. The ancient Iapetus Ocean once lay between the colliding continents. Sedimentary rocks from the bed of this ocean were lifted up, and they have kept the same forms they had in the past. ●

Silurian

The name of the geological period in which this folding occurred.

1. THREE CONTINENTS
The Caledonian orogeny was formed by the collision of three ancient continents: Laurasia, Gondwana, and Baltica. In between them, the Iapetus Ocean floor contained sediments that now form the bedrock of the coast of Wales.

440 MILLION YEARS

395 MILLION YEARS

2. A MOUNTAIN RANGE
The long Caledonian range is seen today in the coasts of England, Greenland, and Scandinavia. Since the tectonic movements that created them have ended, they are being worn away and sculpted by constant erosion.

SANDSTONE

LIMESTONE

Composition

Before mountain ranges were lifted up by the collision of ancient continents, constant erosion of the land had deposited large amounts of sediments along their coasts. These sediments later formed the rock that makes up the folding seen here. As that rock's shape clearly shows, tectonic forces compressed the originally horizontal sediments until they became curved. This phenomenon is seen along Cardigan Bay on the ancient coast of Wales.

SANDSTONE

WALES, UNITED KINGDOM

Latitude: 51° 30′ N

Longitude: 003° 12′ W

Place	Cardigan Bay
Length	40 miles (64 km)
Rock	Sedimentary
Fold	Monoclinal

MUDSTONE

When the Faults Resound

F aults are small breaks that are produced along the Earth's crust. Many, such as the San Andreas fault, which runs through the state of California, can be seen readily. Others, however, are hidden within the crust. When a fault fractures suddenly, an earthquake results. Sometimes fault lines can allow magma from lower layers to break through to the surface at certain points, forming a volcano. ●

Relative Movement Along Fault Lines

Fault borders do not usually form straight lines or right angles; their direction along the surface changes. The angle of vertical inclination is called "dip." The classification of a fault depends on how the fault was formed and on the relative movement of the two plates that form it. When tectonic forces compress the crust horizontally, a break causes one section of the ground to push above the other. In contrast, when the two sides of the fault are under tension (pulled apart), one side of the fault will slip down the slope formed by the other side of the fault.

350 miles
(566 km)

The distance that the opposite sides of the fault have slipped past each other, throughout their history.

① Normal Fault

This fault is the product of horizontal tension. The movement is mostly vertical, with an overlying block (the hanging wall) moving downward relative to an underlying block (the footwall). The fault plane typically has an angle of 60 degrees from the horizontal.

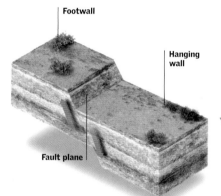

Footwall

Hanging wall

Fault plane

Footwall   **Hanging wall**

② Reverse Fault

This fault is caused by a horizontal force that compresses the ground. A fracture causes one portion of the crust (the hanging wall) to slide over the other (the footwall). Thrust faults (see pages 18-19), are a common form of reverse fault that can extend up to hundreds of miles. However, reverse faults with a dip greater than 45° are usually only a few yards long.

Dip angle

Rodgers Creek
Concord-Green Valley
OAKLAND
Mt. Diablo
SAN FRANCISCO
Greenville
Hayward
Calaveras
San Gregorio

OPPOSITE DIRECTIONS
The northwestward movement of the Pacific Plate and the southeastward movement of the North American Plate cause folds and fissures throughout the region.

PACIFIC OCEAN

③ Oblique-Slip Fault

This fault has horizontal as well as vertical movements. Thus, the relative displacement between the edges of the fault can be diagonal. In the oldest faults, erosion usually smoothes the differences in the surrounding terrain, but in more recent faults, cliffs are formed. Transform faults that displace mid-ocean ridges are a specific example of oblique-slip faults.

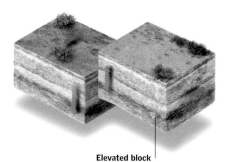

Elevated block

Strike-Slip Fault

In this fault the relative movement of the plates is mainly horizontal, along the Earth's surface, parallel to the direction of the fracture but not parallel to the fault plane. Transform faults between plates are usually of this type. Rather than a single fracture, they are generally made up of a system of smaller fractures, slanted from a centerline and more or less parallel to each other. The system can be several miles wide.

Streambeds Diverted by Tectonic Movement

Through friction and surface cracking, a transform fault creates transverse faults and, at the same time, alters them with its movement. Rivers and streams distorted by the San Andreas fault have three characteristic forms: streambeds with tectonic displacement, diverted streambeds, and streambeds with an orientation that is nearly oblique to the fault.

1

Diverted Streambed
The stream changes course as a result of the break.

2

Displaced Streambed
The streambed looks "broken" along its fault line.

WEST COAST OF THE UNITED STATES

Length of California	**770 miles (1,240 km)**
Length of fault	**800 miles (1,300 km)**
Maximum width of fault	**60 miles (100 km)**
Greatest displacement (1906)	**20 feet (6 m)**

Queen Charlotte Fault

Juan de Fuca Plate

PACIFIC PLATE

San Andreas Fault

NORTH AMERICAN PLATE

San Andreas

OOTWALL   FOOTWALL

East Pacific Ridge

Fault plane

140 years

The average interval between major ruptures that have taken place along the fault. The interval can vary between 20 and 300 years.

PAST AND FUTURE
Some 30 million years ago, the Peninsula of California was west of the present coast of Mexico. Thirty million years from now, it is possible that it may be some distance off the coast of Canada.

Fatal Crack

The great San Andreas fault in the western United States is the backbone of a system of faults. Following the great earthquake that leveled San Francisco in 1906, this system has been studied more than any other on Earth. It is basically a horizontal transform fault that forms the boundary between the Pacific and North American tectonic plates. The system contains many complex lesser faults, and it has a total length of 800 miles (1,300 km). If both plates were able to slide past each other smoothly, no earthquakes would result. However, the borders of the plates are in contact with each other. When the solid rock cannot withstand the growing strain, it breaks and unleashes an earthquake.

Volcanoes

M ount Etna has always been an active volcano, as seen from the references to its activity that have been made throughout history. It could be said that the volcano has not given the beautiful island of Sicily a moment's rest. The Greek philosopher Plato was the first to study Mount Etna. He traveled to Italy especially to see it

MOUNT ETNA
With a height of 10,810 feet (3,295 m), Etna is the largest and most active volcano in Europe.

up close, and he subsequently described how the lava cooled. Today Etna's periodic eruptions continue to draw hundreds of thousands of tourists, who enjoy the spectacular fireworks produced by its red-hot explosions. This phenomenon is visible from the entire east coast of Sicily because of the region's favorable weather conditions and the constant strong winds. ●

Flaming Furnace

olcanoes are among the most powerful manifestations of our planet's dynamic interior. The magma they release at the Earth's surface can cause phenomena that devastate surrounding areas: explosions, enormous flows of molten rock, fire and ash that rain from the sky, floods, and mudslides. Since ancient times, human beings have feared volcanoes, even seeing their smoking craters as an entrance to the underworld. Every volcano has a life cycle, during which it can modify the topography and the climate and after which it becomes extinct. ●

MOUNTAIN-RANGE VOLCANOES

Many volcanoes are caused by phenomena occurring in subduction zones along convergent plate boundaries.

1 When two plates converge, one moves under the other (subduction).

2 The rock melts and forms new magma. Great pressure builds up between the plates.

3 The heat and pressure in the crust force the magma to seep through cracks in the rock and rise to the surface, causing volcanic eruptions.

CRATER

Depression or hollow from which eruptions expel magmatic materials (lava, gas, steam, ash, etc.)

ERUPTION OF LAVA

CLOUD OF ASH

STREAMS OF LAVA

flow down the flanks of the volcano.

PARASITIC VOLCANO

Composite volcanic cones have more than one crater.

SECONDARY CONDUIT

VOLCANIC CONE

Made of layers of igneous rock, formed from previous eruptions. Each lava flow adds a new layer.

LIFE AND DEATH OF A VOLCANO: THE FORMATION OF A CALDERA

1. Explosive eruptions can expel huge quantities of lava, gas, and rock.

2. A void is left in the conduit and in the internal chamber.

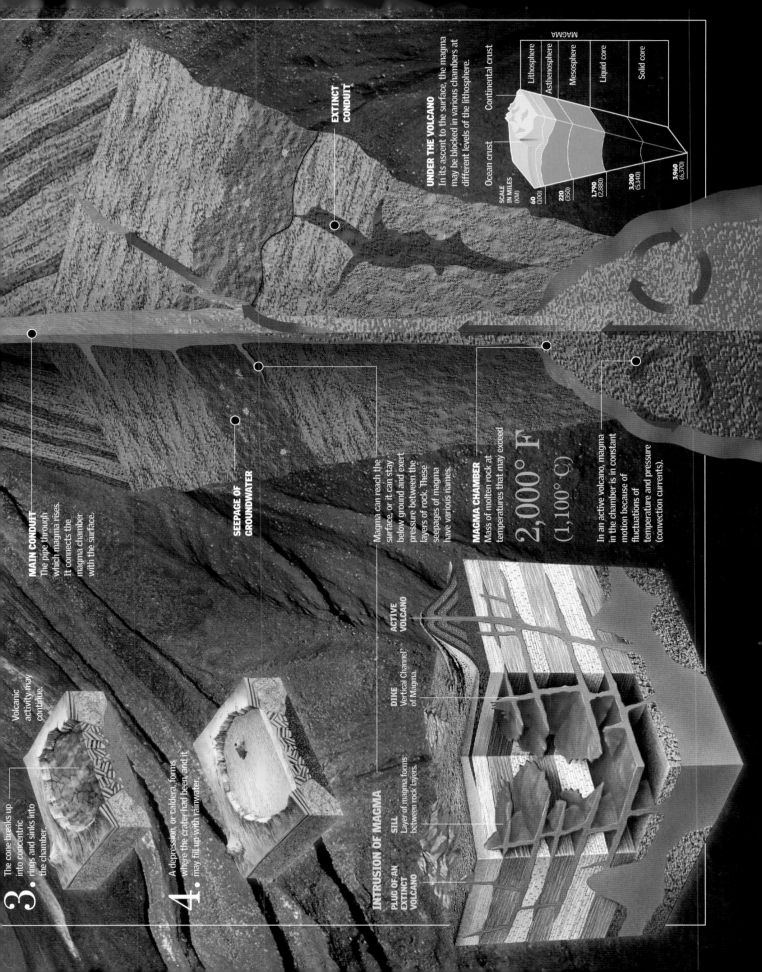

**EXTINCT
CONDUIT**

UNDER THE VOLCANO
In its ascent to the surface, the magma may be blocked in various chambers at different levels of the lithosphere.

MAGMA

Lithosphere

Asthenosphere

Mesosphere

Liquid core

Solid core

Continental crust

Ocean crust

SCALE
IN MILES
(KM)

60
(100)

220
(350)

1,790
(2,880)

3,200
(5,140)

3,960
(6,370)

MAIN CONDUIT
The pipe through which magma rises. It connects the magma chamber with the surface.

**SEEPAGE OF
GROUNDWATER**

Magma can reach the surface, or it can stay below ground and exert pressure between the layers of rock. These seepages of magma have various names.

MAGMA CHAMBER
Mass of molten rock at temperatures that may exceed

2,000° F
(1,100° C)

In an active volcano, magma in the chamber is in constant motion because of fluctuations of temperature and pressure (convection currents).

**ACTIVE
VOLCANO**

DIKE
Vertical Channel of Magma.

INTRUSION OF MAGMA

**PLUG OF AN
EXTINCT
VOLCANO**

SILL
Layer of magma forms between rock layers.

3. Volcanic activity may continue.

The cone breaks up into concentric rings and sinks into the chamber.

4. A depression, or caldera, forms where the crater had been, and it may fill up with rainwater.

Classification

No two volcanoes on Earth are exactly alike, although they have characteristics that permit them to be studied according to six basic types: shield volcanoes, cinder cones, stratovolcanoes, lava cones, fissure volcanoes, and calderas. A volcano's shape depends on its origin, how the eruption began, processes that accompany the volcanic activity, and the degree of danger the volcano poses to life in surrounding areas. ●

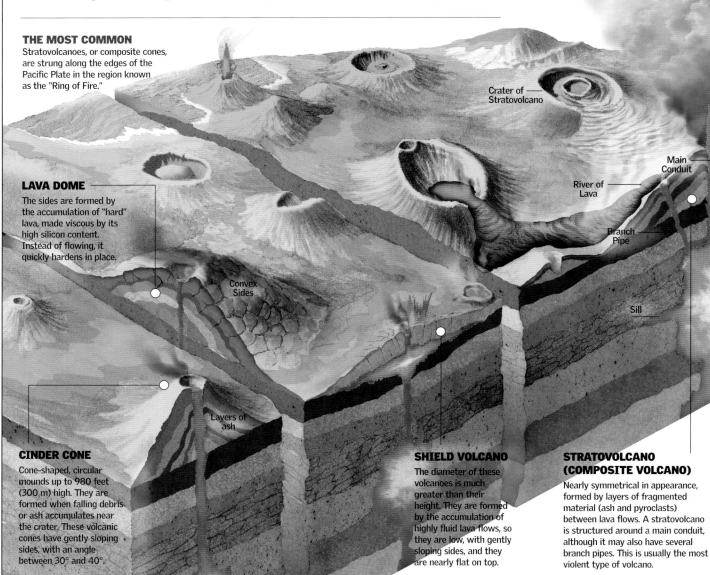

THE MOST COMMON
Stratovolcanoes, or composite cones, are strung along the edges of the Pacific Plate in the region known as the "Ring of Fire."

Crater of Stratovolcano

Main Conduit

River of Lava

Branch Pipe

LAVA DOME
The sides are formed by the accumulation of "hard" lava, made viscous by its high silicon content. Instead of flowing, it quickly hardens in place.

Convex Sides

Sill

Layers of ash

CINDER CONE
Cone-shaped, circular mounds up to 980 feet (300 m) high. They are formed when falling debris or ash accumulates near the crater. These volcanic cones have gently sloping sides, with an angle between 30° and 40°.

SHIELD VOLCANO
The diameter of these volcanoes is much greater than their height. They are formed by the accumulation of highly fluid lava flows, so they are low, with gently sloping sides, and they are nearly flat on top.

STRATOVOLCANO (COMPOSITE VOLCANO)
Nearly symmetrical in appearance, formed by layers of fragmented material (ash and pyroclasts) between lava flows. A stratovolcano is structured around a main conduit, although it may also have several branch pipes. This is usually the most violent type of volcano.

MOUNT ILAMATEPEC
Cinder cone located 45 miles (65 km) west of the capital of El Salvador. Its last recorded eruption was in October 2005.

MOUNT KILAUEA
Shield volcano in Hawaii. One of the most active shield volcanoes on Earth.

MOUNT FUJI
Composite volcano 12,400 feet (3,776 m) high, the highest in Japan. Its last eruption was in 1707.

IGNEOUS INTRUSIONS: A PECULIAR PROFILE

1 **FORMATION OF THE VOLCANIC PLUG**

Extinct volcano

Lava solidifies and forms resistant rock.

2 **INITIAL EROSION**

Erosion of the cone

The plug is not affected.

3 **THE NECK FORMS.**

The surrounding terrain is flat.

The volcanic neck remains.

CHAPEL OF ST. MICHAEL

Built in Le Puy, France, on top of a volcanic neck of hard rock that once sealed the conduit of a volcano. The volcano's cone has long since been worn away by erosion; the lava plug remains.

262
FEET (80 M)
The height of the plug, from base to peak.

Caldera that contains a lake

Plug of extinct volcano

Parasitic Volcano

Lava slope

Formation of new cone

Shock wave

Magma chamber

CALDERA VOLCANO

Large basins, similar to craters but greater than 0.8 mile (1 km) across, are called calderas. They are found at the summit of extinct or inactive volcanoes, and they are typically filled with deep lakes. Some calderas were formed after cataclysmic explosions that completely destroyed the volcano. Others were formed when, after successive eruptions, the empty cone could no longer hold up the walls, which then collapsed.

Dike

FISSURE VOLCANOES

Long, narrow openings found mainly in mid-ocean ridges. They emit enormous amounts of highly fluid material and form wide slopes of stratified basaltic stone. Some, such as that of the Deccan Plateau in India, cover more than 380,000 square miles (1 million sq km).

CALDERA BLANCA

Located on Lanzarote, Canary Islands, in the fissure zone known as the Montañas de Fuego (Fire Mountains).

MAUNA ULU

Fissure volcano, about 5 miles (8 km) from the top of Kilauea (Hawaii). This is one of the most active volcanoes in the central Pacific.

Flash of Fire

A volcanic eruption is a process that can last from a few hours to several decades. Some are devastating, but others are mild. The severity of the eruption depends on the dynamics between the magma, dissolved gas, and rocks within the volcano. The most potent explosions often result from thousands of years of accumulation of magma and gas, as pressure builds up inside the chamber. Other volcanoes, such as Stromboli and Etna, reach an explosive point every few months and have frequent emissions. ●

ASH

LAPILLI

BOMB

HOW IT HAPPENS

3. THE ESCAPE
When the mounting pressure of the magma becomes greater than the materials between the magma and the floor of the volcano's crater can bear, these materials are ejected.

2. IN THE CONDUIT
A solid layer of fragmented materials blocks the magma that contains the volatile gases. As the magma rises and mixes with volatile gases and water vapor, the pockets of gases and steam that form give the magma its explosive power.

1. IN THE CHAMBER
There is a level at which liquefaction takes place and at which rising magma, under pressure, mixes with gases in the ground. The rising currents of magma increase the pressure, hastening the mixing.

4. PYROCLASTIC PRODUCTS
In addition to lava, an eruption can eject solid materials called pyroclasts. Volcanic ash consists of pyroclastic material less than 0.08 inch (2 mm) in size. An explosion can even expel granite blocks.

BOMB	2.5 inches (64 mm) and u
LAPILLI	0.08 to 2.5 inches (2 mm to 64 mm)
ASH	Up to 0.08 inch (2 mm)

5. LAVA FLOWS
On the volcanic island of Hawaii, nonerupting flows of lava abound. Local terms for lava include "aa," viscous lava flows that sweep away sediments, and "pahoehoe," more fluid lava that solidifies in soft waves.

CRATER

Water Vapor

CONDUIT

Gas Particles   Molten Rock

MAGMA CHAMBER

EFFUSIVE ACTIVITY

Mild eruptions with a low frequency of explosions. The lava has a low gas content, and it flows out of openings and fissures.

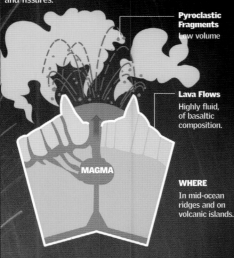

Pyroclastic Fragments
Low volume

Lava Flows
Highly fluid, of basaltic composition.

WHERE
In mid-ocean ridges and on volcanic islands.

MAGMA

EXPLOSIVE ACTIVITY

Comes from the combination of high levels of gas with relatively viscous lava, which can produce pyroclasts and build up great pressure. Different types of explosions are distinguished based on their size and volume. The greatest explosions can raise ash into a column several miles high.

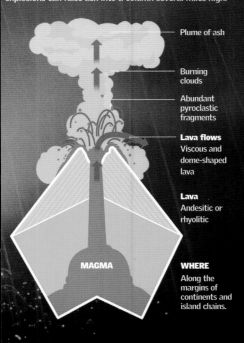

Plume of ash

Burning clouds

Abundant pyroclastic fragments

Lava flows
Viscous and dome-shaped lava

Lava
Andesitic or rhyolitic

WHERE
Along the margins of continents and island chains.

MAGMA

TYPES OF EFFUSIVE ERUPTION

Dome Low, like a shield volcano, with a single opening

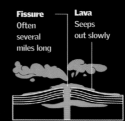

Large, Frequent Lava Flows

Fissure Often several miles long

Lava Seeps out slowly

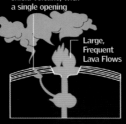

HAWAIIAN

Volcanoes such as Mauna Loa and Kilauea expel large amounts of basaltic lava with a low gas content, so their eruptions are very mild. They sometimes emit vertical streams of bright lava ("fountains of fire") that can reach up to 330 feet (100 m) in height.

FISSURE

Typical in ocean rift zones, fissures are also found on the sides of composite cones such as Etna (Italy) or near shield volcanoes (Hawaii). The greatest eruption of this type was that of Laki, Iceland, in 1783: 2.9 cubic miles (12 cu km) of lava was expelled from a crack 16 miles (25 km) long.

TYPES OF EXPLOSIVE ERUPTION

Cloud can reach above **82,000 feet (25 km).**

The column can reach a height of **49,000 feet (15 km)**

Cloud of burning material from about **330 to 3,300 feet (100-1,000 m)** high

Lava flow

Burning cloud moving down the slope

Lava plug

STROMBOLIAN

The volcano Stromboli in Sicily, Italy, gave its name to these high-frequency eruptions. The relatively low volume of expelled pyroclasts allows these eruptions to occur approximately every five years.

VULCANIAN

Named after Vulcano in Sicily. As eruptions eject more material and become more explosive, they become less frequent. The 1985 eruption of Nevado del Ruiz expelled tens of thousands of cubic yards of lava and ash.

VESUVIAN

Also called Plinian, the most violent explosions raise columns of smoke and ash that can reach into the stratosphere and last up to two years, as in the case of Krakatoa (1883).

PELEAN

A plug of lava blocks the crater and diverts the column to one side after a large explosion. As with Mt. Pelée in 1902, the pyroclastic flow and lava are violently expelled down the slope in a burning cloud that sweeps away everything in its path.

FROM OUTER SPACE

A photo of the eruption of Mt. Augustine in Alaska, taken by the Landsat 5 satellite hours after the March 27, 1986, eruption.

SMOKE COLUMN

7 Miles
(11.5 Km) HIGH

Volcanic ash

Snow and ice

Lava flow

LAVA FLOW **MT. KILAUEA, HAWAII**   LAKE OF LAVA **MAKA-O-PUHL, HAWAII**   COOLED LAVA (PAHOEHOE) **MT. KILAUEA, HAWAII**

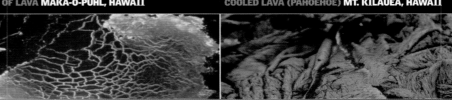

Mount St. Helens

W ithin the territory of the United States, active volcanoes are not limited to exotic regions such as Alaska or Hawaii. One of the most explosive volcanoes in North America is in Washington state. Mount St. Helens, after a long period of calm, had an eruption of ash and vapor on May 18, 1980. The effects were devastating: 57 people were killed, and lava flows destroyed trees over an area of 232 square miles (600 sq km). The lake overflowed, causing mudslides that destroyed houses and roads. The area will need a century to recover.

PRECOLLAPSE SUMMIT

GLACIER

NEW DOME

OLD DOME (1980-86)

GLACIER TONGUE

CONE

The Top
Like the cork in a bottle of champagne, the top of the mountain burst off because of pressure from the magma.

OLYMPIA WASHINGTON STATE

Type of Volcano	Stratovolcano
Size of Base	5.9 mi (9.5 km)
Type of Activity	Explosive
Type of Eruption	Plinian
Most Recent Eruptions	1980, 1998, 2004
Fatalities	57

Warning Signs

Two months before the great explosion, Mount St. Helens gave several warning signs: a series of seismic movements, small explosions, and a swelling of the mountain's north slope, caused by magma rising toward the surface. Finally on May 18, an earthquake caused a landslide that carried away the top of the volcano. Later, several collapses at the base of the column caused numerous pyroclastic flows with temperatures of nearly 1,300° F (700° C).

1.
00:00

SWELLING
The uninterrupted flow of magma toward the volcano's surface caused the north slope of the mountain to swell, and later collapse in an avalanche.

Influx of magma.

Unchanged profile.

Precollapse swelling.

Secondary dome of earlier rocks.

9,680 feet
(2,950 m)

-1,315 feet
(-401 m)

In the eruption Mount St. Helens lost its conical stratovolcano shape and became a caldera.

8,363 feet
(2,549 m)

BEFORE THE ERUPTION
The symmetrical cone, surrounded by forest and prairies, was admired as the American Fuji. The eruption left a horseshoe-shaped caldera, surrounded by devastation.

DURING THE EXPLOSION
The energy released was the equivalent of 500 nuclear bombs. The top of the mountain flew off like the cap of a shaken bottle of soda.

232 SQUARE MILES
600 sq km

SURFACE DESTRUCTION
The effects were devastating: 250 houses, 47 bridges, rail lines, and 190 miles (300 km) of highway were lost.

8 miles
13 km

Pulverized and incinerated by the force of the lava and the pyroclastic flow. Temperatures rose above 1,110° F (600° C).

15 miles
24 km

Range of the shock wave from the pyroclastic flow. The heat and ash left acres of forest completely destroyed.

The Forest
Burned trees covered with ash, several miles from the volcano

2.
00:40

PRESSURE ON THE NORTH SLOPE
The swelling of the mountain was no doubt caused by the first eruption, almost two months before the final explosion.

Graben: Depression caused by movement in the Earth's crust

Blocked Crater

Having no escape route, the magma exerts pressure sideways and breaks through the north slope.

3.
00:50

INITIAL ERUPTIONS
The north slope gave way to the great pressure of the magma in an explosive eruption. The lava traveled 16 miles (25 km) at 246 feet (75 m) per second.

Side block of the cone

The crater exploded.

The side block gave way, causing a powerful pyroclastic flow.

4.
00:60

EXPLOSION AND VERTICAL COLLAPSE
At the foot of the volcano, a valley 640 feet (195 m) deep was buried in volcanic material. Over 10 million trees were destroyed.

A vertical column of smoke and ash rose 12 miles (19 km) high.

Profile before the collapse

Profile after the collapse

Krakatoa

In early 1883, Krakatoa was just one of many volcanic islands on Earth. It was located in the Straits of Sundra, between Java and Sumatra in the Dutch East Indies, now known as Indonesia. It had an area of 10.8 square miles (28 sq km) and a central peak with a height of 2,690 feet (820 m). In August 1883, the volcano exploded, and the island was shattered in the largest natural explosion in history. ●

The Island That Exploded

Krakatoa was near the subduction zone between the Indo-Australian and Eurasian plates. The island's inhabitants were unconcerned about the volcano because the most recent previous eruption had been in 1681. Some even thought the volcano was extinct. On the morning of Aug. 27, 1883, the island exploded. The explosion was heard as far away as Madagascar. The sky was darkened, and the tsunamis that followed the explosion were up to 130 feet (40 m) high.

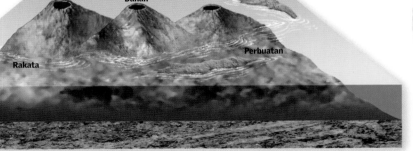

Danan

Rakata

Perbuatan

BEFORE
In May the volcano began showing signs in the form of small quakes and spouting vapor, smoke, and ash. None of this served to warn of the terrible explosion to come, and some even took trips to see the volcano's "pyrotechnics."

DURING
At 5:30 a.m. the island burst from the accumulated pressure, opening a crater 820 feet (250 m) deep. Water immediately rushed in, causing a gigantic tsunami.

34 miles (55 km)

The height of the column of ash.

130 feet (40 m)

The height of the tsunami waves, which traveled at 700 miles per hour (1,120 km/h).

KRAKATOA

Latitude 6° 06´ S
Longitude 105° 25´ E

Surface Area	10.8 square miles (28 sq km)
Remaining Surface Area	3 square miles (8 sq km)
Range of the Explosion	2,900 miles (4,600 km)
Range of Debris	1,550 miles (2,500 km)
Tsunami Victims	36,000

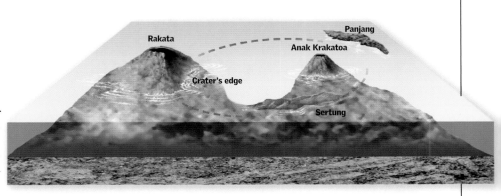

Rakata

Panjang

Anak Krakatoa

Crater's edge

Sertung

PYROCLASTICS
The pyroclastic flows were so violent that, according to the descriptions of sailors, they reached up to 37 miles (80 km) from the island.

3 ## AFTER
A crater nearly 4 miles (6.4 km) in diameter was left where the volcano had been. About 1927, new volcanic activity was observed in the area. In 1930, a cone emerged. Anak Krakatoa ("daughter of Krakatoa") appeared in 1952; it grows at a rate of nearly 15 feet (4.5 m) per year.

FRACTION
Two thirds of the island was destroyed, and only a part of Rakata survived the explosion.

Aftereffects

The ash released into the atmosphere left enough particles suspended in the air to give the Moon a blue tinge for years afterward. The Earth's average temperature also decreased.

Long-Term Effects

WATER LEVEL
The water level fluctuated as far away as the English Channel.

PRESSURE WAVE
The atmospheric pressure wave went around the world seven times.

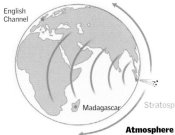

English Channel

Madagascar

Stratosphere

Atmosphere
The ash expelled by the explosion lingered for years

500

MEGATONS
The energy released, equivalent to 25,000 atomic bombs such as the one dropped on Hiroshima.

Aftermath of Fury

When a volcano becomes active and explodes, it sets in motion a chain of events beyond the mere danger of the burning lava that flows down its slopes. Gas and ash are expelled into the atmosphere and affect the local climate. At times they interfere with the global climate, with more devastating effects. The overflow of lakes can also cause mudslides called lahars, which bury whole cities. In coastal areas, lahars can cause tsunamis. ●

SNOW

LAVA

VOLCANO

MUD

LAVA FLOWS

In volcanoes with calderas, low-viscosity lava can flow without erupting, as with the Laki fissures in 1783. Low-viscosity lava drips with the consistency of clear honey. Viscous lava is thick and sticky, like crystallized honey.

LAVA IN VOLCANO NATIONAL PARK, HAWAII

CINDER CONE

Cone with walls of hardened lava.

As the lava flows upward, the cone explodes.

MOLDS OF TREES

Burned tree underneath cooled lava.

The petrified mold forms a minivolcano.

LAVA TUBES

Outer layer of hardened lava.

Inside, the lava stays hot and fluid.

RESCUE IN ARMERO, COLOMBIA

Mudslide after the eruption of the volcano Nevado del Ruiz. A rescue worker helps a boy trapped in a lahar.

MUDSLIDES OR LAHARS

Rain mixed with snow and melted by the heat, along with tremors and overflowing lakes, can cause mudslides called "lahars." These can be even more destructive than the eruption itself, destroying everything in their path as they flow downhill. They occur frequently on high volcanoes that have glaciers on their summit.

ARMERO FROM ABOVE

On Nov. 13, 1985, the city of Armero, Colombia, was devastated by mudslides from the eruption of the volcano Nevado del Ruiz.

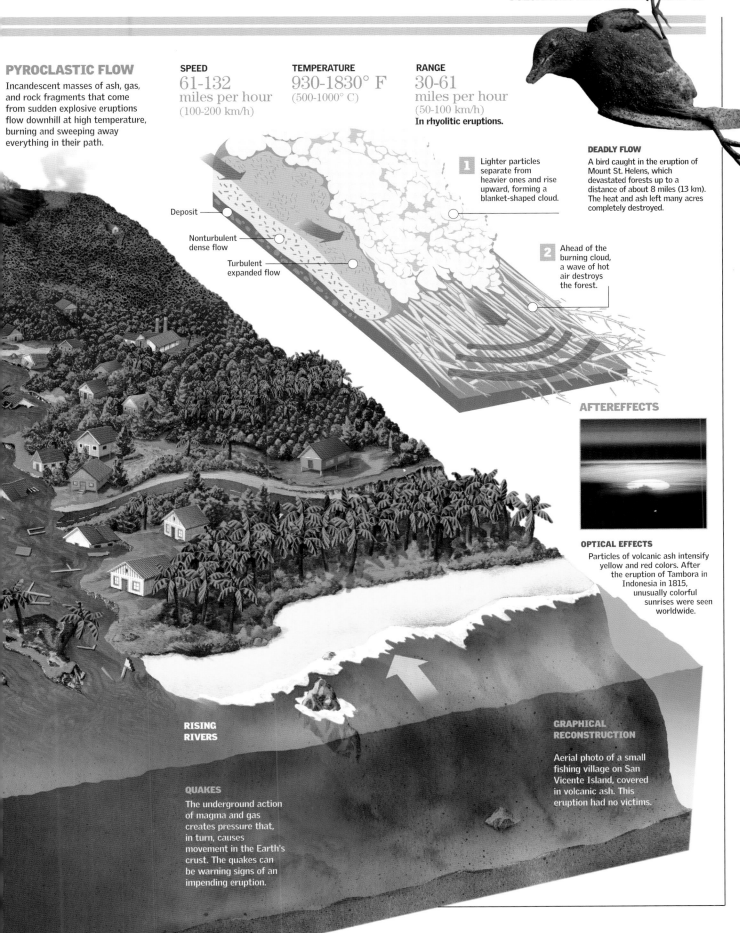

PYROCLASTIC FLOW

Incandescent masses of ash, gas, and rock fragments that come from sudden explosive eruptions flow downhill at high temperature, burning and sweeping away everything in their path.

SPEED
61-132
miles per hour
(100-200 km/h)

TEMPERATURE
930-1830° F
(500-1000° C)

RANGE
30-61
miles per hour
(50-100 km/h)
In rhyolitic eruptions.

1 Lighter particles separate from heavier ones and rise upward, forming a blanket-shaped cloud.

Deposit

Nonturbulent dense flow

Turbulent expanded flow

DEADLY FLOW

A bird caught in the eruption of Mount St. Helens, which devastated forests up to a distance of about 8 miles (13 km). The heat and ash left many acres completely destroyed.

2 Ahead of the burning cloud, a wave of hot air destroys the forest.

AFTEREFFECTS

OPTICAL EFFECTS

Particles of volcanic ash intensify yellow and red colors. After the eruption of Tambora in Indonesia in 1815, unusually colorful sunrises were seen worldwide.

RISING RIVERS

GRAPHICAL RECONSTRUCTION

Aerial photo of a small fishing village on San Vicente Island, covered in volcanic ash. This eruption had no victims.

QUAKES

The underground action of magma and gas creates pressure that, in turn, causes movement in the Earth's crust. The quakes can be warning signs of an impending eruption.

Jets of Water

Geysers are intermittent spurts of hot water that can shoot up dozens of yards into the sky. Geysers form in the few regions of the planet with favorable hydrogeology, where the energy of past volcanic activity has left water trapped in subterranean rocks. Days or weeks may pass between eruptions. Most of these spectacular phenomena are found in Yellowstone National Park (U.S.) and in northern New Zealand. ●

The eruptive cycle

5. THE CYCLE REPEATS

When the water pressure in the chambers is relieved, the spurt of water abates, and the cycle repeats. Water builds up again in cracks of the rock and in permeable layers.

On average, a geyser can expel up to

7,900 gallons
(30,000 l) OF WATER

4. SPURTING SPRAY

The water spurts out of the cone at irregular intervals. The lapse between spurts depends on the time it takes for the chambers to fill up with water, come to a boil, and produce steam.

The average height reached of the spurt of water is about

148 feet
(45 m)

Streams of water and steam

PRINCIPAL GEOTHERMAL FIELDS

There are some 1,000 geysers worldwide, and 50 percent are in Yellowstone National Park (U.S.).

Kamchatka (Russia)

Great Geysir (Iceland)

North Island (New Zealand)

Umnak Island (U.S.)

Steamboat Springs/ Beowawe (U.S.)

El Tatio (Chile)

YELLOWSTONE (U.S.)

GRAND PRISMATIC SPRING

This spring, in Yellowstone National Park, is the largest hot spring in the United States and the third largest in the world. It measures 246 by 377 feet (75 by 115 m), and it emits about 530 gallons (2,000 l) of water per minute. It has a unique color: red mixed with yellow and green.

In the middle of the spring, the mineral water is 200° F (93° C), and it cools gradually toward the edges.

Path

377 feet (115 m)

DISCHARGE

530 gallons
(2,000 l)

OF WATER PER MINUTE

RECORD HEIGHT

In 1904, New Zealand's Waimangu geyser (now inactive) emitted a record-setting spurt of water. In 1903, four tourists lost their lives when they unknowingly came too close to the geyser.

1,450 ft
(442 m)

1,500 ft
(457 m)

TALLEST U.S. BUILDING

RECORD HEIGHT

OTHER POSTVOLCANIC ACTIVITY

FUMAROLE

This is a place where there is a constant emission of water vapor because the temperature of the magma is above 212° F (100° C).

Water vapor

Hot water

SOLFATARA

The thermal layers emit sulfur and sulfurous anhydride.

Sulfurous gases

Steam

MUD BASIN

These basins produce their own mud; sulfuric acid corrodes the rocks on the surface and creates a mud-filled hollow.

Mud, clay, mineral deposits, and water

Hot water

Steam Energy

In Iceland, geothermic steam is used not only in thermal spas but also to power turbines that generate most of the country's electricity.

MINERAL SPRINGS

Their water contains many minerals, known since antiquity for their curative properties. Among other substances, they include sodium, potassium, calcium, magnesium, silicon oxide, chlorine, sulfates (SO4), and carbonates (HCO3). They are very helpful for rheumatic illnesses.

TERRACES
These are shallow, quickly drying pools with stair-step sides.

CRATER

CHIMNEY

CONE

SECONDARY CONDUIT

MAIN VENT

RESERVOIR OR CHAMBER

Geyser with multiple chambers

HEAT SOURCE
Magma between 2 and 6 miles (3-10 km) deep, at 930-1,110° F (500-600° C).

3. BURSTING FORTH

The water rises by convection and spurts out the main vent to the chimney or cone. The deepest water becomes steam and explodes outward.

CONVECTION FORCES

This is a phenomenon equivalent to boiling water.

A
Water cools and sinks back to the interior, where it is reheated.

B
Bubbles of hot gas rise to the surface and give off their heat.

2. MOUNTING PRESSURE

The underground chambers fill with water, steam, and gas at high temperatures, and these are then expelled through secondary conduits to the main vent.

Temperatures up to
194° F
(90° C)

1. HEATED WATER

Thousands of years after the eruption of a volcano, the area beneath it is still hot. The heat rising from the magma chambers warms water that filters down from the soil. In the subsoil, the water can reach temperatures of up to 518° F (270° C), but pressure from cooler water above keeps it from boiling.

MORPHOLOGY OF THE CHAMBERS

The heat of a magma chamber warms water in the cavity, the chamber fills, and the water rises to the surface. The pressure in the cavity is released, and the water suddenly boils and spurts out

Grand Fountain (Yellowstone)

Old Faithful (Yellowstone)

Great Geysir (Iceland)

Great Fountain (Yellowstone)

Narcissus (Yellowstone)

Round Geyser (Yellowstone)

Rings of Coral

In the middle of the ocean, in the tropics, there are round, ring-shaped islands called atolls. They are formed from coral reefs that grew along the sides of ancient volcanoes that are now submerged. As the coral grows, it forms a barrier of reefs that surround the island like a fort. How does the process work? Gradually, volcanic islands sink, and the reefs around them form a barrier. Finally, the volcano is completely submerged; no longer visible, it is replaced by an atoll.

FORMATION OF AN ATOLL

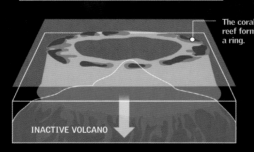

1. **THE BEGINNING OF AN ATOLL.** The undersea flanks of an extinct volcano are colonized by corals, which continue to grow.

VOLCANIC CONE

CORAL REEF

INACTIVE VOLCANO

2. **THE CORALS GAIN GROUND.** As the surrounding reef settles and continues to expand, it becomes a barrier reef that surrounds the summit of the ancient volcano, now inactive.

CORAL REEF

INACTIVE VOLCANO

3. **THE ATOLL SOLIDIFIES.** Eventually the island will be completely covered and will sink below the water, leaving a ring of growing coral with a shallow lagoon in the middle.

The coral reef forms a ring.

INACTIVE VOLCANO

TAKARAYAN

CORAL REEF

BEACH

INNER REEF

LIMESTON

REEF LEVELS

CREST
Barrier that protects the shore from waves. Deep grooves and tunnels let seawater inside the reef.

FACE
Branching corals grow here, though colonies can break loose because of the steep slope.

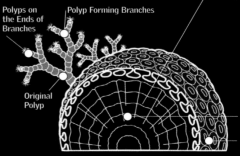

WHAT ARE CORALS?

Corals are formed from the exoskeletons of a group of Cnidarian species. These marine invertebrates have a sexual phase, called a medusa, and an asexual phase, called a polyp. The polyps secrete an outer skeleton composed of calcium carbonate, and they live in symbiosis with one-celled algae.

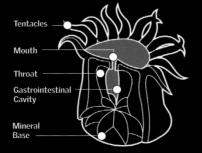

HARD CORAL POLYP

Tentacles

Mouth

Throat

Gastrointestinal Cavity

Mineral Base

BRANCHING CORAL

Polyps on the Ends of Branches

Polyp Forming Branches

Original Polyp

COMPACT CORAL

Original polyp formation (dead)

Layer of live polyps

ATOLLS AND VOLCANIC ISLANDS AROUND THE WORLD

Coral reefs are found in the world's oceans, usually between the Tropic of Cancer and the Tropic of Capricorn.

TROPIC OF CANCER

TROPIC OF CAPRICORN

KIRIBATI

OPTIMAL CONDITIONS

Coral is mainly found in the photic zone (less than 165 feet [50 m] deep), where sunlight reaches the bottom and provides sufficient energy. For reefs to grow, the water temperature should be between

68° and 82° F (20-28° C).

BUOTA

TEMOTU

RAWANNAWI

M A R A K E I

ANTAI

TEKUANGA

INNER LAKE

NORAUEA

N

Scale in miles (km)	
0 (0)	0.6 (1)
0.3 (0.5)	

TEROKEA

Country	**Republic of Kiribati**
Ocean	**North Pacific**
Archipelago	**Gilbert Islands**
Surface area	**10.8 square miles (28 sq km)**
Altitude	**6.9 ft (2.1 m)**

LEGEND

● Town   ✪ Capital

HAWAIIAN ARCHIPELAGO

Nihau Kauai   Oahu   Molokai   Maui

Lanai

Kahoolave   Hawaii

FORMATION OF A VOLCANIC ISLAND

A Volcanoes form when magma rises from deep within the Earth. Thousands of volcanoes form on the seafloor, and many emerge from the sea and form the base of islands.

B When a plate of the crust moves over a hot spot, a volcano begins to erupt and an island is born.

Plate movement

Molokai	Kohoolave	Lanai	Maui	Hawaii
1,476 ft (450 m)	3,369 ft (1,027 m)	3,369 ft (1,027 m)	10,023 ft (3,055 m)	13,799 ft (4,206 m)

Frozen Flame

It is known as the land of ice and fire. Under Iceland's frozen surface there smolders a volcanic fire that at times breaks free and causes disasters. The island is located over a hot spot on the Central Atlantic Ridge. In this area the ocean bed is expanding, and large quantities of lava flow from vents, fissures, and craters.

1/5

of all the lava that has emerged on the Earth's surface since 1500 has come from Iceland.

ICELAND

Latitude 64° 6' N
Longitude -21° 54' E

Surface Area	39,768 sq miles (103,000 sq km)
Population	293,577
Population density	1 per sq mile (2.8 per sq km)
Area of lakes	1,064 sq miles (2,757 sq km)
Glaciers	4,603 sq miles (11,922 sq km)

ENERGY

The islanders use geothermal (steam) energy from volcanoes and geysers for heat, hot water, and electric energy.

SNAEFELLS
LYSUHOLL

REYKJAVIK

The capital of Iceland is the northernmost capital in the world.

PRESTAHNU

REYKJAVIK ○

HENGILL

VATNAFJOL

REYKJANES

Split Down the Middle

Part of Iceland rests on the North American Plate, which is drifting westward. The rest of Iceland is on the Eurasian Plate, drifting eastward. As tectonic forces pull on the plates, the island is slowly splitting in two and forming a fault. The edges of the two plates are marked by gorges and cliffs. Thus, the ocean bed is growing at the surface.

North American Plate

Eurasian Plate

Average Annual Expansion: 0.4 inch (1 cm)

REYKJAVIK

Mid-Atlantic Ridge

Atlantic Ocean

The magma that emerges at the surface comes from a series of central volcanoes separated by fissures.

Birth of an Island

On Nov. 15, 1963, an undersea volcanic eruption off the southern coast of Iceland gave rise to the island of Surtsey, the newest landmass on the planet. The eruption began with a large column of ash and smoke. Later, heat and pressure deep within the Earth pushed part of the Mid-Atlantic Ridge to the surface. The island kept growing for several months, and today it has a surface area of 1.0 square mile (2.6 sq km). The island was named after Surtur, a fire giant from Icelandic mythology.

SURTSEY

RIFT ZONE

If the rift zone that crosses the island from southwest to north were cut in two, different ages of the Earth would be revealed according to the segment being analyzed. For example, the rock 60 miles (100 km) from the rift is six million years old.

6 miles (10 km) per millions of years ←   6 miles (10 km) per millions of years →
100 75 50 25   25 50 75 100

Depth in miles (km)

3(5)
6(10)

4 mya
6 mya
8 mya
10 mya

600°C
800°C
1,000°C

Rift Zone

This volcano has been very active throughout history. Of its 29 active periods, the most recent was in 1984.

THEISTARE

hot spot   ▲ KRAFLA

Lake Myvatn

FREMRINAM

Crater of 1,640 feet (500 m). The caldera measures **6 miles (10 km)** across.

ASKJA

Lake Viti (Hell in Icelandic), Krafla Volcano

HOFSJOKUL

KERLINGAR

▲ BARDARBUNGA

GLACIAL CAP OF VATNAJÖKULL

▲ GRIMSVÖTN

ERUPTION UNDER THE ICE

In 1996 a fissure opened up between Grimsvötn and Bardarbunga. The lava made a hole 590 feet (180 m) deep in the ice and released a column of ash and steam. The eruption lasted 13 days.

HEKLA Its largest eruption was in 1104.

LAKI

The largest eruption of lava in history occurred in 1783: the ashes reached China.

TINDFJALL

KATLA   ▲ VESTMANNA

1 The first eruptions were caused by the interaction of magma and water. The explosions were infrequent, and rocks were thrown only a few yards from the volcano.

2 Repeated eruptions expelled vapor and ash into the air, forming a column over 6 miles (10 km) high. The island was formed from volcanic blocks and masses of lava.

3 The entire process lasted three-and-a-half years. Over 0.25 cubic mile (1 cu km) of lava and ash was expelled, with only 9 percent of it appearing above sea level.

Study and Prevention

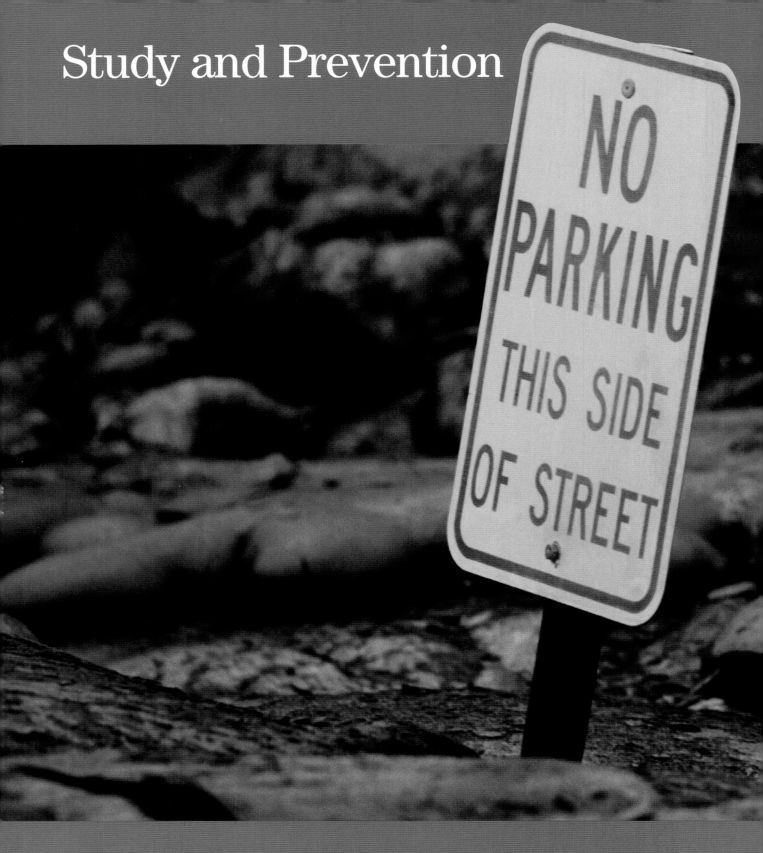

Large eruptions often give warning signs months in advance. These signs consist of any observable manifestation on the exterior of the Earth's crust. They may include emissions of steam, gases, or ash and rising temperatures in the lake that typically forms in the crater. This is why volcanic seismology is

INCANDESCENT ROCK
A river of lava from Mt. Kilauea flows constantly, forming surface wrinkles that deform under the lightest step.

considered one of the most useful tools for protecting nearby towns. Several seismic recording stations are typically placed around the cone of an active volcano. Among other things, the readings scientists get give them a clear view of the varying depths of the volcano's tremors-extremely important data for estimating the probability of a major eruption. ●

Latent Danger

Some locations have a greater propensity for volcanic activity. Most of these areas are found where tectonic plates meet, whether they are approaching or moving away from each other. The largest concentration of volcanoes is found in a region of the Pacific known as the "Ring of Fire." Volcanoes are also found in the Mediterranean Sea, in Africa, and in the Atlantic Ocean.

Arctic Ocean

AVACHINSKY
Russia
This is a young, active cone inside an old caldera, on the Kamchatka peninsula.

NOVARUPTA
Alaska, U.S.
It is in the Valley of Ten Thousand Smokes.

MOUNT ST. HELENS
Washington, U.S.
It had an unexpected, violent eruption in 1980.

The Pacific "Ring of Fire"
Formed by the edges of the Pacific tectonic plate, where most of the world's volcanoes are found.

FUJIYAMA
Japan
This sacred mountain is the country's largest volcano.

ASIA

PINATUBO
Philippines
In 1991 it had the second most violent eruption of the 20th century.

50 Volcanoes
Indonesia has the highest concentration of volcanoes in the world. Java alone has 50 active volcanoes.

MAUNA LOA
Hawaii, U.S.
The largest active volcano on Earth is rooted in the ocean floor and takes up nearly half of the island.

KILAUEA
Hawaii, U.S.
The most active shield volcano, its lava flows have covered more than 40 square miles (100 sq km) since 1983.

OCEANIA

Pacific Ocean

KRAKATOA
Indonesia
Its 1883 eruption destroyed an entire island.

TAMBORA
Indonesia
In 1815 it produced 35 cubic miles (150 cu km) of ash. It was the largest recorded eruption in human history.

EAST EPI
Vanuatu
This is an undersea caldera with slow eruptions lasting for months.

Subduction
Most volcanoes in the western United States were formed by subduction of the Pacific Plate.

Indian Ocean

AUSTRALIAN PLATE

PACIFIC PLATE

The tallest

These are found in the middle of the Andes range, which forms part of the Pacific Ring of Fire. They were most active 10,000 years ago, and many are now extinct or dampened by fumarolic action.

OJOS DEL SALADO	LLULLAILLACO	TIPAS	INCAHUASI	SAJAMA	MAUNA LOA
Chile/Argentina	Chile/Argentina	Argentina	Chile/Argentina	Bolivia	**Hawaii**
22,595 ft	22,110 ft	21,850 ft	21,720 ft	21,460 ft	Shield volcano
(6,887 m)	(6,739 m)	(6,660 m)	(6,621 m)	(6,542 m)	13,680 feet (4,170 m) above sea level.

CALDERA

The "top five" list changes when the volcanoes are measured from the base rather than from their altitude above sea level.

SEA LEVEL

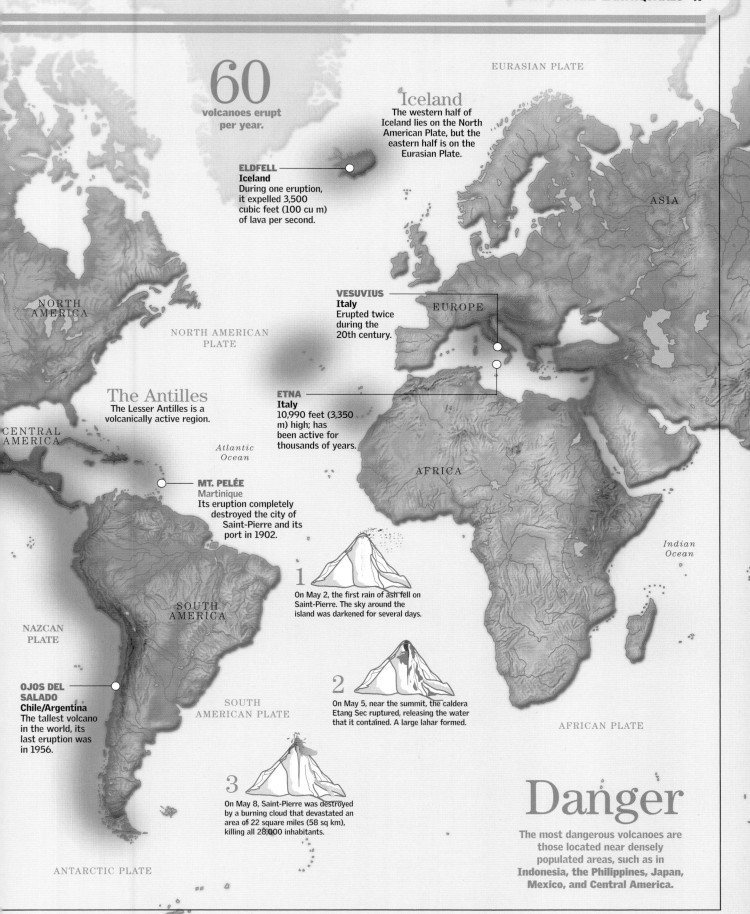

60
volcanoes erupt
per year.

Iceland
The western half of
Iceland lies on the North
American Plate, but the
eastern half is on the
Eurasian Plate.

EURASIAN PLATE

ASIA

ELDFELL
Iceland
During one eruption,
it expelled 3,500
cubic feet (100 cu m)
of lava per second.

VESUVIUS
Italy
Erupted twice
during the
20th century.

EUROPE

NORTH
AMERICA

NORTH AMERICAN
PLATE

The Antilles
The Lesser Antilles is a
volcanically active region.

ETNA
Italy
10,990 feet (3,350
m) high; has
been active for
thousands of years.

*Atlantic
Ocean*

AFRICA

CENTRAL
AMERICA

MT. PELÉE
Martinique
Its eruption completely
destroyed the city of
Saint-Pierre and its
port in 1902.

*Indian
Ocean*

1

On May 2, the first rain of ash fell on
Saint-Pierre. The sky around the
island was darkened for several days.

SOUTH
AMERICA

NAZCAN
PLATE

2

On May 5, near the summit, the caldera
Etang Sec ruptured, releasing the water
that it contained. A large lahar formed.

AFRICAN PLATE

**OJOS DEL
SALADO**
Chile/Argentina
The tallest volcano
in the world, its
last eruption was
in 1956.

SOUTH
AMERICAN PLATE

3

On May 8, Saint-Pierre was destroyed
by a burning cloud that devastated an
area of 22 square miles (58 sq km),
killing all 28,000 inhabitants.

Danger

The most dangerous volcanoes are
those located near densely
populated areas, such as in
Indonesia, the Philippines, Japan,
Mexico, and Central America.

ANTARCTIC PLATE

Increasing Knowledge

Volcanology is the scientific study of volcanoes. Volcanologists study eruptions from airplanes and satellites, and they film volcanic activity from far off. However, to study the inner workings of a volcano up close, they must scale near-vertical cliffs and face the dangers of lava, gas, and mudslides. Only then can they take samples and set up equipment to detect tremors and sounds.

Field Measurements

Monitoring a volcano includes gathering and analyzing samples and measuring various phenomena. Seismic movements, varying compositions of gases, deformations in the rock, and changes in electromagnetic fields induced by the movement of underground magma can all provide clues to predict volcanic activity.

FUMAROLE

B

146. °C

PRONTO ™ Tc
DIGITAL THERMOMETER

ON OFF

TYPE K THERMOCOUPLE
-80 °C TO 800 °C

THERMO ELECTRIC

Made in the EEC

Thermopar

A   B

Magma

GPS Receiver

LAVA TEMPERATURE
is measured with a thermometer called a Thermopar; glass thermometers would melt from the heat. Temperatures of water and of nearby rocks are other variables to take into account.

TILTOMETERS
These are placed on the slopes of a volcano to record soil changes that precede an eruption. Points are determined (A-B) to monitor how pressure from the magma deforms the surface between them.

GLOBAL POSITIONING SYSTEM (GPS)
Movements in the magma cause hundreds of cracks in the cone. A GPS system records images continuously and analyzes the deformation over a period of time.

SAMPLING OF VOLCANIC GASES

Gas and water vapor dissolved in magma provides the energy that powers eruptions. Visible emissions, such as sulfur and steam, are measured, as are invisible gases. Analyzing the gases' composition makes it possible to predict the beginning and end of an eruption.

GAS MASK

VACUUM TUBE

TITANIUM TUBE

GAS ESCAPING FROM CRACK

Volcanologists Taking Gas Samples from a Fumarole on Lipari Island, Italy

Portable seismograph

TEMBLORS OR EARTHQUAKES

Portable seismographs are used to detect movements in the ground within 6 miles (10 km) of the volcano that is being studied. These tremors can give clues about the movements of the magma.

Taking measurements of the crater

THE SIZE OF THE CRATER

The widening of the crater caused by volcanic activity and the growth of the solid lava domes are measured. This growth implies certain risks for the proximity of an eruption.

Lahar detector

HYDROLOGICAL MONITORING

Mudslides, or lahars, can bury large areas. Monitoring the volume of water in the area makes it possible to alert and evacuate the population when the amount of water passes critical points.

Lava sample

LAVA COLLECTION

The study of lava can determine its mineral composition and its origin. Lava deposits are also analyzed because the history of a volcano's eruptions can give clues about a future eruption.

Preparations for Disaster

Volcanic eruptions are dangerous to surrounding populations for two basic reasons. One danger is posed by the volcanic material that flows down the sides of the volcano (lava flows and mudslides), and the other danger is from the volcano's pyroclastic material, especially ash. Ash fallout can bury entire cities. Experts have developed an effective series of prevention and safety measures for people living in volcanic areas. These measures greatly reduce the highest risks. ●

12 miles
(20 km)

Before an Eruption

It is best to get informed about safety measures, evacuation routes, safe areas, and alarm systems before a volcanic eruption. Other safety measures include stocking up on nonperishable food, obtaining gas masks and potable water, and checking the load-bearing capacity of roofs.

Do not carry more than

44 pounds
(20 kg) OF PROVISIONS.

RIVERS AND STREAMS
Large volumes of water pose a threat of mudslides. Avoid these areas.

MAIN ROUTES
These usually cross low-lying areas. They can be a potential path for flows of lava or mud.

BRIDGES
When possible, do not use for your evacuation route because they might collapse.

Evacuation of Nearby Areas

In the immediate area (within 12 miles [20 km]) of the volcano, evacuation is the only possible safety measure. Returning home will be possible only when permission is given. Keep in mind that it takes a long time for life to return to normal after an evacuation.

MEDICAL PRECAUTIONS
Keep a first-aid kit and essential medications on hand, and keep vaccinations up to date.

SHUT OFF UTILITIES
Before leaving a house, shut off the electricity, gas, and water. Tape doors and windows shut.

HIGHER ELEVATIONS
These are the preferred sites for evacuations from volcanic eruptions. High ground is safe from lahars and lava flows, and if there is shelter there, it is also safe from rains of ash.

PROVISIONS
Water and food are indispensable, especially if you evacuate the area on your own.

CIVIL DEFENSE
Follow all recommendations, be alert to official information, and do not spread rumors.

LAHARS AND PYROCLASTIC FLOWS
Lahars (mudflows) can form from rainwater or melting snow. Volcanic danger zones often have strategies to divert rivers and reduce the volume of water in dams and reservoirs.

12 miles
(20 km)
Considered to be the critical distance from a volcano in emergency relief efforts.

WIND AND RAIN
Wind is a risk factor that spreads volatile ash over a large area so that settlements at a distance greater than 60 miles (100 km) can be affected. The greatest danger posed by falling ash is that it can mix with rain falling on the roofs of houses and form a heavy mass that will collapse the buildings.

60 miles
(100 km)

Areas of Falling Ash

Most of the population lives outside the volcano's range, but ash from an eruption can become highly volatile and fall over wide areas. Wind can carry ash to other areas, so the best preventive efforts are focused on warning people about what to do in case of falling ash.

ALTERNATIVE ROUTES
Roads running through higher elevations are preferred because they cannot be reached by lava and mudflows.

WATER TANK
Roof-mounted water tanks should be disconnected and covered until the roof has been cleared of ashes.

AT HOME
It is best to stay indoors during an ashfall. One of the main precautions is to provide for potable drinking water, because the usual water supply will be interrupted because of pollution risks, especially if the water supply comes from lakes or rivers in the area.

DOORS AND WINDOWS
It is best to always leave doors and windows shut tightly, as airtight as possible, for as long as the ashfall continues.

ASH ON THE ROOF
Ash should be removed immediately (before it rains) so the roof does not collapse.

AIR CONDITIONING
Air conditioners and large clothes dryers should not be used during an eruption.

MASKS
Use masks and special ash-protective clothing when outdoors.

STAY CALM
To breathe, cover your face with a handkerchief soaked in water and vinegar.

INFORMATION
Listen to the radio at all times.

AVOID DRIVING
If you must drive, do so slowly and turn on your headlights. It is best to leave the car parked in an enclosed space or under cover.

DO NOT WASH WITH WATER.
After the ashfall, washing with water will form a sticky and heavy paste that will be very hard to remove.

CHILDREN
If children are at school, do not go to pick them up: they will be safe there.

Buried in One Day

A t noon on Aug. 24, AD 79, Mount Vesuvius erupted near the coast of Naples Bay. The Roman cities of Pompeii and Herculaneum were completely buried in ashes and pyroclasts, in what would become one of the worst natural tragedies of ancient times. Many details from that day have reached us thanks to the narrative of Pliny the Younger. His well-known description of the eruption column as "shaped like a pine" caused this type of eruption to be named after him: a "Plinian eruption." ●

10 A.M.

1 **AN ALMOST NORMAL DAY**
Tremors and earthquakes had been felt in the city for four days. Hanging lamps swayed, furniture moved, and some door frames had even cracked. Because these things happened about once a year without any consequences, the inhabitants of Pompeii continued with their normal lives. The public forum was filled with people. The festivities of Isis were celebrated in the temple of Apollo.

POMPEII'S FORUM
This was the political, religious, and commercial heart of the city. Every day the forum was alive with Pompeii's citizens, as it was on August 24.

1 P.M.

2 **THE ERUPTION.**
Suddenly Vesuvius spewed out a huge column of smoke, lava, and ash that formed a pyroclastic flow moving toward Pompeii. People ran in all directions seeking refuge in houses. The roughness of the sea made escape by water impossible.

RICHES
Several precious objects such as this gold bracelet have been unearthed.

The Violent Awakening

Mount Vesuvius had been inactive for more than 800 years, until the pressure that had accumulated inside produced its explosion in the year 79. Most of the deaths during this tragedy were originally blamed on the ash that buried parts of the neighboring settlements (Herculaneum and Stabiae, as well as Pompeii). Now, though, the eruption is believed to have produced the typical "burning clouds" of a Plinian eruption: Flames of incandescent ash and gases were expelled at high speeds by the eruptive pressure. Suspended moist particles charged the air with electricity, causing an intense electric storm, whose flashes of lightning would have been the only source of light under the ashfall. Since then Vesuvius has had a dozen other important eruptions. The worst killed 4,000 people in 1631. The first volcanology observatory in the world was installed at Vesuvius in 1841.

POMPEII, ITALY

Latitude 40° 49' N
Longitude 14° 26' E

Distance from Vesuvius	6 miles (10 km)
Population in the year AD 79	20,000 people
Current population	27,000 people
Ash dispersion (79)	60 miles (100 km) (SE)
Last Eruption of Vesuvius	1944

23 feet
(7 m)
The maximum depth of the ashes.

9 P.M.

3 A TWO-DAY NIGHT
The tongues of lava from the volcano were seen better at night. The next morning the Sun's light could not be seen through the ash cloud. Pliny's narrative mentions a constant rain of pyroclasts, continuing on the following morning, and emissions of sulfuric gases that killed many people. Many sought shelter on the beaches. Only on August 26 did the ashfall begin to disperse.

SEQUENCE OF THE ERUPTION

For more than 20 hours (the time the eruption lasted), the ash column rose and then fell on the surrounding area.

1 After the first explosion, the column of smoke began its vertical climb. The wind blew it toward the southeast.

2 The cloud spread nearly 60 miles (100 km) from side to side, and ash fell on the city for a whole day.

3 By 7:30 A.M. on August 25, the pyroclastic flows reached Pompeii. These flows are estimated to have reached temperatures of 1,022° F (550° C).

RAIN OF STONES
Moments after the eruption, incandescent pumice stones fell from the sky.

In the House of the Faun

Objects and human bodies were found under Pompeii's ashes, preserved in the position in which the disaster surprised them. These valuable testimonies to the past have made possible the reconstruction of daily life in ancient Rome. The House of the Faun was one of the most luxurious villas in Pompeii.

Slaves worked in the kitchens, and there were utensils similar to those we use today.

The funnel-shaped roofs were used to collect rainwater.

Clients seeking protection or favors were received in the atrium or central patio.

Slave couples could meet only in the gardens.

BRONZE FAUN
The name given to the house is from this statue found in the villa's atrium. The faun was considered a wild deity, with the ability to predict the future.

Tiles decorated with the flora and fauna of the Nile.

This merchant home was the largest in Pompeii, with 32,290 square feet (3,000 sq m).

Good Eating and Drinking

Romans were more than fond of feasts. A dinner for the whole family, which normally began at four in the afternoon, could last for more than four hours. Meals were sumptuous affairs, and no one left until completely satisfied. Pompeii's wine was famous throughout Rome. Kept in pitchers, it was always served watered down. The Romans sometimes added flavorings, commonly including honey and pepper.

As in That Moment

In 1709, some of Pompeii's artifacts were found buried under volcanic ash, and that started a treasure hunt. It was not until 1864, though, that reconstruction and conservation of materials began with the work of Giuseppe Fiorelli. The exhibits that are the most fascinating to people who visit Pompeii's ruins today (about two million people every year) are his reconstructions of the bodies.

In a Pompeii Bar

There were several types of food and drink establishments in Pompeii, from food vendors in the streets to luxury services. These food places served many different social purposes but acted primarily as places for businessmen to meet. They were run mostly by slaves, men as well as women.

1 THE CATASTROPHE
Several corpses were covered by volcanic ash, which had accumulated in layers and later hardened. The bodies had decomposed, but their forms were molded in the volcanic rock.

2 RECONSTRUCTION
The work of Fiorelli was to fill these natural "tombs" (ash molds) with plaster. When the plaster hardened, the surrounding layers of ash were removed, leaving the outlines or molds of the bodies.

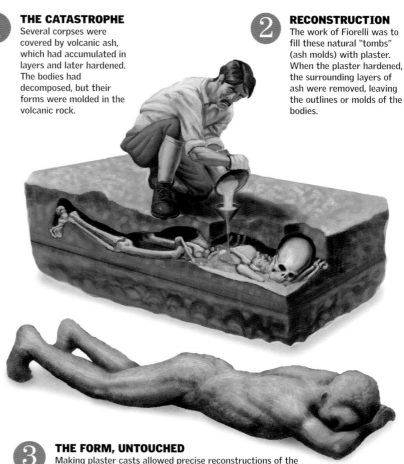

3 THE FORM, UNTOUCHED
Making plaster casts allowed precise reconstructions of the people's postures at the time of the disaster, and we have been able to learn details such as the hairstyles and dress of these people. Animal forms and other organic objects have also been reconstructed. Today the use of resins and silicones makes it possible to obtain even greater detail.

THE THERMOPOLIUM
The typical bar had a long marble tabletop with embedded containers in which food could be kept warm.

200
was the approximate number of places of this type in the city.

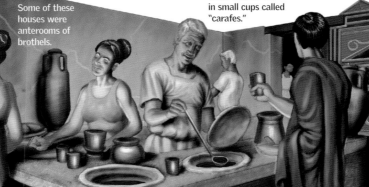

Some of these houses were anterooms of brothels.

Wine was served in small cups called "carafes."

Food included large quantities of nuts, olives, bread, cheese, and onions.

There were frescoes on the walls showing obscene images and pictures of drunken customers.

Historic Eruptions

The lava falls and flows, sweeping away everything in its path. This happens in a slow, uninterrupted way, and the lava destroys entire cities, towns, and forests and claims thousands of human lives. One of the most famous examples was the eruption of Mount Vesuvius in AD 79, which wiped out two cities and two cultures, those of Pompeii and Herculaneum. In the 20th century, the eruption of Mount Pelée destroyed the city of Saint-Pierre in Martinique in a few minutes and instantly killed almost its entire population. Volcanic activity also seems to be closely related to changes in climate.

AD 79

MOUNT VESUVIUS
Naples, Italy

Volume of ejected ash in cubic feet (cu m)	141,000 (4,000)
Victims	2,200
Characteristics	Active

The cities of Pompeii and Herculaneum were destroyed in AD 79 when Mount Vesuvius erupted. Until that day, it was not known that the mountain was a volcano because it had been inactive for over 300 years. This was one of the first eruptions to be recorded: Pliny the Younger stated in one of his manuscripts that he had seen how the mountain exploded. He described the gas and ash cloud rising above Vesuvius and how thick, hot lava fell. Many people died because they inhaled the poisonous gases.

Volcanoes and Climate

There is a strongly supported theory that relates climate changes to volcanic eruptions. The idea of linking the two phenomena is based on the fact that explosive eruptions spew huge amounts of gases and fine particles high into the stratosphere, where they spread around the Earth and remain for years. The volcanic material blocks a portion of solar radiation, reducing air temperatures around the world. Perhaps the most notable cold period related to volcanic activity was the one that followed the eruption of Tambora in 1815. Some areas of North America and Europe had an especially harsh winter.

KALAPANA. After the Kilauea volcano (Hawaii) erupted in 1991, a lava flow advanced on the city, covering everything in its path.

1783

LAKI VOLCANO
Iceland

Volume of ejected ash in cubic feet (cu m)	490 billion (14 billion)
Victims	10,000
Characteristics	Very active

In spite of the fact that the eruptions are related to conic forms, most of the volcanic material comes out through fractures in the crust, called "fissures." The fissure eruptions of Laki were the greatest in Iceland; they created more than 20 vents in a distance of 15 miles (25 km). The gases ruined grasslands and killed livestock. The subsequent famine took the lives of 10,000 people.

1815

TAMBORA VOLCANO
Indonesia

Volume of ejected ash in cubic feet (cu m)	100 billion (3 billion)
Victims	10,000
Characteristics	Stratovolcano

After giving off fumes for seven months, Tambora erupted, and the ensuing catastrophe was felt around the globe. The ash cloud expanded to more than 370 miles (600 km) away from the epicenter of the eruption, and it was so thick that it hid the Sun for two days. The ashfall covered an area of 193,051 square miles (500,000 sq km). It is considered to be the most destructive volcanic explosion that ever took place. More than 10,000 people died during the eruption, and 82,000 died of illness and starvation after the eruption.

1883

KRAKATOA VOLCANO
Java, Indonesia

Volume of ejected ash in cubic feet (cu m)	**670 billion (19 billion)**
Victims	**36,000**
Characteristics	**Active**

Even though Krakatoa began to announce its forthcoming eruption with clouds of vapor and smoke, these signs, instead of preventing a disaster, became a tourist attraction. When the explosion took place, it destroyed two thirds of the island. Stones shot from the volcano reached a height of 34 miles (55 km)- beyond the stratosphere. A crater 4 miles (6.4 km) in diameter opened a chasm 820 feet (250 m) deep. Land and islands were swept bare.

1902

MOUNT PELÉE
Martinique, Antilles

Volume of ejected ash in cubic feet (cu m)	**No figures available**
Victims	**30,000**
Characteristics	**Stratovolcano**

A burning cloud and a thick mass of ash and hot lava were shot from this small volcano that completely destroyed the port city of Saint-Pierre. Most striking is the fact that this destruction took place in only a few minutes. The energy released was so great that trees were uprooted. Almost the entire population died, and only three people survived, one of them because he was trapped in the city jail.

1973

ELDFELL VOLCANO
Heimaey Islands, Iceland

Volume of ejected ash in cubic feet (cu m)	**No figures available**
Victims	**0**
Characteristics	**656 feet (200 m)**

The lava advanced, and it appeared that it would take everything in its path. Volcanologists decided that Heimaey Island, south of Iceland, should be evacuated. But a physics professor proposed watering the lava with seawater to solidify or harden it. Forty-seven pumps were used, and, after three months and 6.5 million tons (6 million metric tons) of water, the lava was stopped, and the port was saved. The eruption began on January 23 and ended on June 28.

1980

MOUNT ST. HELENS
State of Washington, U.S.

Volume of ejected ash in cubic feet (cu m)	**35 billion (1 billion)**
Victims	**57**
Characteristics	**Active**

Also known as the Mount Fuji of the American continent. During the 1980 explosion, 1,315 feet (401 m) of the mountain's top gave way through a fault on its side. A few minutes after the volcano began its eruption, rivers of lava flowed down its sides, carrying away the trees, houses, and bridges in their path. The eruption destroyed whole forests, and the volcanic debris devastated entire communities.

1944

MOUNT VESUVIUS
Naples, Italy

Volume of ejected ash in cubic feet (cu m)	**No figures available**
Victims	**2,000**
Characteristics	**End of a cycle**

With this last activity, the Vesuvius volcano ended the cycle of eruptions it began in 1631. This explosion, along with the previous one in 1906, caused severe material damage. The eruptions were responsible for more than 2,000 deaths from avalanches and lava bombs. Additionally, the 1944 eruption took place during World War II and caused as much damage as the eruption at the beginning of the 20th century had, because it flooded Somma, Atrio de Cevallo, Massa, and San Sebastiano.

1982

EL CHICHÓN VOLCANO
Mexico

Volume of ejected ash in cubic feet (cu m)	**No figures available**
Victims	**2,000**
Characteristics	**Active**

On Sunday, March 28, after 100 years of inactivity, this volcano became active again and unleashed an eruption on April 4. The eruption caused the deaths of about 2,000 people who lived in the surrounding area, and it destroyed nine settlements. It was the worst volcanic disaster in Mexico's history.

Earthquakes

E arthquakes shake the ground in all directions, even though the effects of a quake depend on the magnitude, depth, and distance from its point of origin. Often the waves are so strong that the Earth buckles, causing the collapse of houses and buildings, as happened in Loma Prieta. In mountainous regions earthquakes

can be followed by landslides and mudslides, whereas in the oceans, tsunamis may form; these walls of water strike the coast with enough force to destroy whole cities, as occurred in Indonesia in December 2004. Thailand recorded the highest number of tourist deaths, and 80 percent of tourist areas were destroyed. ●

Deep Rupture

E arthquakes take place because tectonic plates are in constant motion, and therefore they collide with, slide past, and in some cases even slip on top of each other. The Earth's crust does not give outward signs of all the movement within it. Rather energy builds up from these movements within its rocks until the tension is more than the rock can bear. At this point the energy is released at the weakest parts of the crust. This causes the ground to move suddenly, unleashing an earthquake. ●

1 FORESHOCK
Small tremor that can anticipate an earthquake by days or even years. It could be strong enough to move a parked car.

2 AFTERSHOCK
New seismic movement that can take place after an earthquake. At times it can be even more destructive than the earthquake itself.

EARTHQUAKES PER YEAR

30 Seconds
The time lapse between each tremor of the Earth's crust

MAGNITUDE	QUANTITY
8 or Greater	1
7 to 7.9	18
6 to 6.9	120
5 to 5.9	800
4 to 4.9	6,200
3 to 3.9	49,000

EPICENTER
Point on the Earth's surface located directly above the focus.

HYPOCENTER OR FOCUS
Point of rupture, where the disturbance originates. Can be up to 435 miles (700 km) below the surface.

SOUTHERN ALP

ALPINE FAULT

7.05

7.65 ●
Richter

PLAIN

FAULT PLANE
Usually curves rather than following a straight line. This irregularity causes the tectonic plates to collide, which leads to earthquakes as the plates move.

FOLDS
These result from tension that accumulates between tectonic plates. Earthquakes release part of the tension energy generated by orogenic folds.

ORIGIN OF AN EARTHQUAKE

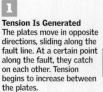

1
Tension Is Generated
The plates move in opposite directions, sliding along the fault line. At a certain point along the fault, they catch on each other. Tension begins to increase between the plates.

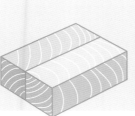

2
Tension Versus Resistance
Because the force of displacement is still active even when the plates are not moving, the tension grows. Rock layers near the boundary are distorted and crack.

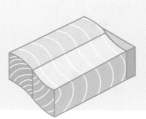

3
Earthquake
When the rock's resistance is overcome, it breaks and suddenly shifts, causing an earthquake typical of a transform-fault boundary.

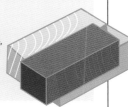

③ EARTHQUAKE
The main movement or tremor lasts a few seconds, after which some alterations become visible in the terrain near the epicenter.

NEW ZEALAND
Latitude 42° S
Longitude 174° E

Surface area	**103,737 square miles (268,680 sq km)**
Population	**4,137,000**
Population density	**35 people per square mile (13.63 people per sq km)**
Earthquakes per year (>4.0)	**60-100**
Total earthquakes per year	**14,000**

SOUTH ISLAND

Riverbeds follow a curved path because of movement along the fault line.

LAKE TEKAPO

6.10

SEISMIC WAVES transmit the force of the earthquake over great distances in a characteristic back-and-forth movement. Their intensity decreases with distance.

15 miles (25 km)
Average depth of the Earth's crust below the island.

NORTH ISLAND

Potential earthquake zone

Australian Plate

Alpine fault

SOUTH ISLAND

Pacific Plate

ALPINE FAULT IN NEW ZEALAND
As seen in the cross-section, South Island is divided by a large fault that changes the direction of subduction, depending on the area. To the north the Pacific Plate is sinking under the Indo-Australian Plate at an average rate of 1.7 inches (4.4 cm) per year. To the south, the Indo-Australian Plate is sinking 1.4 inches (3.8 cm) per year under the Pacific Plate.

FUTURE DEFORMATION OF THE ISLAND

To the west there is a plain that has traveled nearly 310 miles (500 km) to the north in the past 20 million years.

2 million years

4 million years

Elastic Waves

Focus
Vibrations travel outward from the focus, shaking the rock.

S eismic energy is a wave phenomenon, similar to the effect of a stone dropped into a pool of water. Seismic waves radiate out in all directions from the earthquake's hypocenter, or focus. The waves travel faster through hard rock and more slowly through loose sediment and through water. The forces produced by these waves can be broken down into simpler wave types to study their effects. ●

2.2 miles per second (3.6 km/s)
S waves are 1.7 times as slow as P waves.

They travel only through solids. They cause splitting motions that do not affect liquids. Their direction of travel is perpendicular to the direction of travel.

Different Types of Waves

There are basically two types of waves: body waves and surface waves. The body waves travel inside the Earth and transmit foreshocks that have little destructive power. They are divided into primary (P) waves and secondary (S) waves. Surface waves travel only along the Earth's surface, but, because of the tremors they produce in all directions, they cause the most destruction.

➤ Direction of seismic waves

➤ Vibration of rock particles

3.7 miles per second (6 km/s)
Typical Speed of P Waves in the Crust.

P waves travel through all types of material, and the waves themselves move in the direction of travel.

Primary Waves

High-speed waves that travel in straight lines, compressing and stretching solids and liquids they pass through.

SPEED IN DIFFERENT MATERIALS

MATERIAL	Granite	Basalt	Limestone	Sandstone	Water
Wave speed in feet per second (m/s)	17,000 (5,200)	21,000 (6,400)	7,900 (2,400)	11,500 (3,500)	4,800 (1,450)

The ground is **compressed and stretched** by turns along the path of wave propagation.

Surface Waves

appear on the surface after the P and S waves reach the epicenter. Having a lower frequency, surface waves have a greater effect on solids, which makes them more destructive.

1.9 miles per second (3.2 km/s)
Speed of surface waves in the same medium.

These waves travel only along the surface, at 90 percent of the speed of S waves.

LOVE WAVES

These move like horizontal S waves, trapped at the surface, but they are somewhat slower and make cuts parallel to their direction.

RAYLEIGH WAVES

These waves spread with an up-and-down motion, similar to ocean waves, causing fractures perpendicular to their travel by stretching the ground.

The soil **is moved** to both sides.

The ground is moved in an **elliptical** pattern.

The soil **is moved** to both sides, **perpendicular** to the wave's path of motion.

Secondary Waves

Body waves that shake the rock up and down and side to side as they move.

SPEED IN DIFFERENT MATERIALS

MATERIAL	Granite	Basalt	Limestone	Sandstone
Wave speed in feet per second (m/s)	9,800 (3,000)	1,500 (3,200)	4,430 (1,350)	7,050 (2,150)

TRAJECTORY OF P AND S WAVES

The Earth's outer core acts as a barrier to S waves, blocking them from reaching any point that forms an angle greater than 105° from the epicenter. P waves are transmitted farther through the core, but they may be diverted later on. Thus they are detected at points that form an angle of greater than 140° from the epicenter.

Primary (P) Waves
Secondary (S) Waves

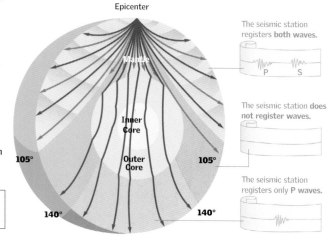

Epicenter

Mantle

Inner Core

Outer Core

105°
105°

140°
140°

The seismic station registers **both waves**.

P   S

The seismic station does **not register waves**.

The seismic station registers only P waves.

Types of Earthquakes

Although earthquakes generally cause all types of waves, some kinds of waves may predominate. This fact leads to a classification that depends on whether vertical or horizontal vibration causes the most movement. The depth of the epicenter can also affect its destructiveness.

BASED ON TYPE OF MOVEMENT

Trepidatory
Located near the epicenter, where the vertical component of the movement is greater than the horizontal.

Oscillatory
When a wave reaches soft soil, the horizontal movement is amplified, and the movement is said to be oscillating.

BASED ON FOCUS DEPTH

Earthquakes originate at points between 3 and 430 miles (5 and 700 km) underground. Ninety percent originate in the first 62 miles (100 km). Those originating between 43 and 190 miles (70 and 300 km) are considered intermediate. Superficial earthquakes (often of higher magnitude) occur above that level, and deep-focus earthquakes occur below it.

Superficial   **43 miles (70 km)**
Intermediate   **190 miles (300 km)**
Deep focus   **430 miles (700 km)**

OAKLAND, CALIFORNIA

Latitude 37° 46′ N

Longitude 122° 13′ W

Surface area of state	156,100 square miles (404,298 sq km)
Range of Damages	68 miles (110 km)
State population	36,132,147
Earthquakes per year (>4.0)	15-20
Earthquake victims	63
Magnitude on Richter scale	7.1

Bursts of tension

Seismic energy can even be compared to the power of nuclear bombs. In addition, the interaction between seismic waves and soil materials also causes a series of physical phenomena that can intensify its destructive capacity. An example of such an effect occurred when a section of an interstate highway plummeted to the ground during the 1989 earthquake in Loma Prieta, California.

THE HIGHWAY

Each soil type responds differently to an earthquake. The figure shows how the same quake can produce waves of different strengths in different soils of different composition. The 0.86-mile (1.4-km) collapsed section of Interstate 880 was built on the mud of San Francisco Bay.

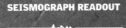

SEISMOGRAPH READOUT

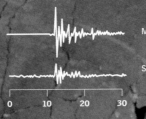

Rock

Mud

Sand

0   10   20   30

FACTORS THAT DETERMINE THE EFFECTS OF AN EARTHQUAKE

INTRINSIC
Magnitude
Type of wave
Depth

GEOLOGIC
Distance
Wave direction
Topography
Groundwater saturation

SOCIAL
Quality of construction
Preparedness of the population
Time of day

4 = 1,000 tons

Magnitude   Energy released by TNT

The energy released by an earthquake of magnitude 4 on the Richter scale is equal to the energy released by a low-power atomic bomb.

7 = 32 million tons

Magnitude   Energy released by TNT

The energy released by an earthquake of magnitude 7 on the Richter scale, such as the 1995 earthquake in Hyogo-Ken Nanbu, Japan, is equal to the energy released by a high-powered thermonuclear bomb (32 megatons).

12 = 160 quadrillion tons

Magnitude   Energy released by TNT

The energy released by a hypothetical earthquake of magnitude 12 on the Richter scale (the greatest known earthquake was 9.5) would be equal to the energy released if the Earth were to split in half.

Liquefaction

Seismic tremors apply a force to muddy or water-saturated soils, filling the empty spaces between grains of sand. Solid particles become suspended in the liquid, the soil loses its load-bearing capacity, and buildings sink as if the ground were quicksand. That displaces some of the water, which rises to the surface.

DIRECT AND INDIRECT EFFECTS

Direct effects are felt in the fault zone and are rarely seen at the surface. Indirect effects stem from the spread of seismic waves. In the Kobe earthquake, the fault caused a fissure in the island of Awaji up to 10 feet (3 m) deep. The indirect effects had to do with liquefaction.

Water liquefies the soil

Gravity

Building

Sinking

1 The soil is compact, even though it contains water.

2 During the earthquake, the water causes the solid particles to shake.

3 Solid structures sink, and water rises.

Measuring an Earthquake

E arthquakes can be measured in terms of force, duration, and location. Many scientific instruments and comparative scales have been developed to take these measurements. Seismographs measure all three parameters. The Richter scale describes the force or intensity of an earthquake. Naturally, the destruction caused by earthquakes can be measured in many other ways: numbers of people left injured, dead, or homeless, damage and reconstruction costs, government and business expenditures, insurance costs, school days lost, and in many more ways. ●

CHARLES RICHTER
American seismologist (1900-85) who developed the scale of magnitude that bears his name.

Intensity

Concept of the destruction caused by an earthquake.

Modified Mercalli Scale

Between 1883 and 1902, this Italian volcanologist developed a scale to measure the intensity of earthquakes. It originally had 10 points based on the observation of the effects of seismic activity; it was later modified to 12. The first few levels consist of barely perceptible sensations. The highest levels apply to the destruction of buildings. This scale is widely used to compare levels of damage among different regions and socioeconomic conditions.

Richter Scale

In 1935, seismologist Charles Richter designed a scale to measure the amplitude of the largest waves registered by seismographs. An important feature of this scale is that the levels increase exponentially. Each point on the scale represents 10 times the movement and 30 times the energy of the point below it. Temblors of magnitude 2 or less are not perceptible to humans. This scale is the most widely used in the world because it can be used to compare the strength of earthquakes apart from their effects.

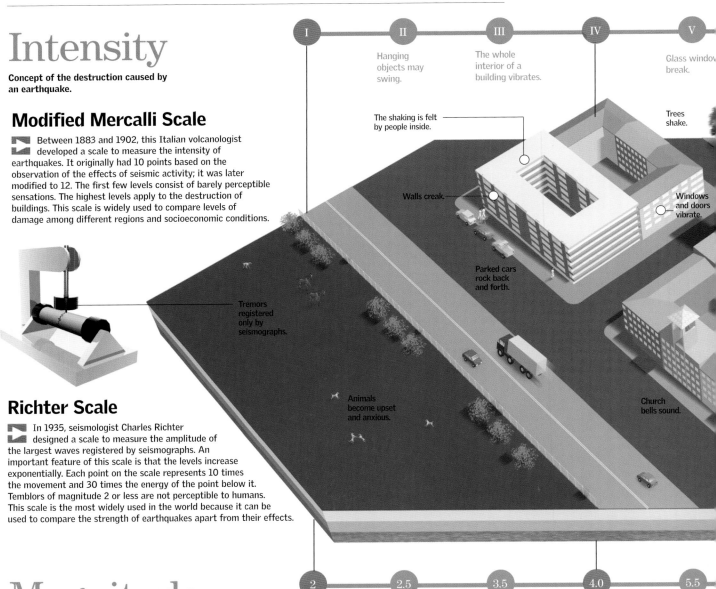

I	II	III	IV	V
	Hanging objects may swing.	The whole interior of a building vibrates.		Glass windows break.

The shaking is felt by people inside.

Trees shake.

Walls creak.

Windows and doors vibrate.

Tremors registered only by seismographs.

Parked cars rock back and forth.

Church bells sound.

Animals become upset and anxious.

Magnitude

The energy released in a seismic event.

2	2.5	3.5	4.0	5.5
Registered only by seismographs.	Very few people feel the tremor.	The tremor is felt. Only minor damages.	Most people perceive the quake.	Some buildings are lightly damaged.

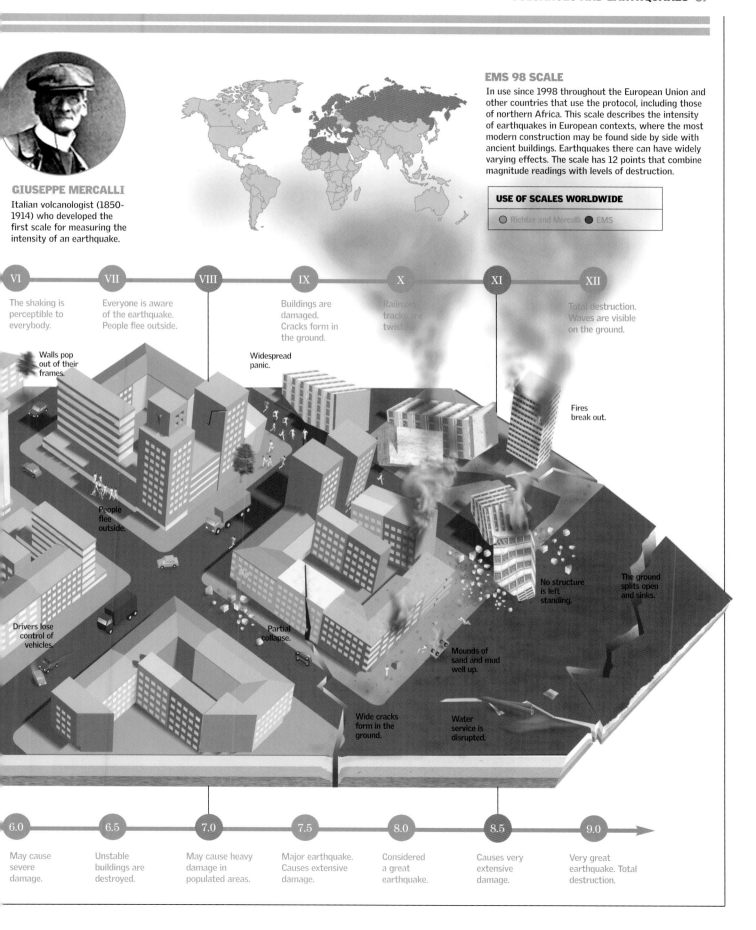

GIUSEPPE MERCALLI

Italian volcanologist (1850-1914) who developed the first scale for measuring the intensity of an earthquake.

EMS 98 SCALE

In use since 1998 throughout the European Union and other countries that use the protocol, including those of northern Africa. This scale describes the intensity of earthquakes in European contexts, where the most modern construction may be found side by side with ancient buildings. Earthquakes there can have widely varying effects. The scale has 12 points that combine magnitude readings with levels of destruction.

USE OF SCALES WORLDWIDE

○ Richter and Mercalli ● EMS

VI The shaking is perceptible to everybody.

VII Everyone is aware of the earthquake. People flee outside.

VIII Buildings are damaged. Cracks form in the ground.

X Railroad tracks are twisted.

XII Total destruction. Waves are visible on the ground.

Walls pop out of their frames.

Widespread panic.

Fires break out.

People flee outside.

No structure is left standing.

The ground splits open and sinks.

Drivers lose control of vehicles.

Partial collapse.

Mounds of sand and mud well up.

Wide cracks form in the ground.

Water service is disrupted.

6.0 May cause severe damage.

6.5 Unstable buildings are destroyed.

7.0 May cause heavy damage in populated areas.

7.5 Major earthquake. Causes extensive damage.

8.0 Considered a great earthquake.

8.5 Causes very extensive damage.

9.0 Very great earthquake. Total destruction.

Violent Seas

A large earthquake or volcanic eruption can cause a tsunami, which means "wave in the harbor" in Japanese. Tsunamis travel very fast, up to 500 miles per hour (800 km/h). On reaching shallow water, they decrease in speed but increase in height. A tsunami can become a wall of water more than 33 feet (10 m) high on approaching the shore. The height depends partly on the shape of the beach and the depth of coastal waters. If the wave reaches dry land, it can inundate vast areas and cause considerable damage. A 1960 earthquake off the coast of Chile caused a tsunami that swept away communities along 500 miles (800 km) of the coast of South America. Twenty-two hours later the waves reached the coast of Japan, where they damaged coastal towns. ●

The word tsunami comes from Japanese

TSU NAMI
Harbor Wave

How It Happens

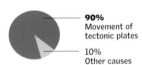
A tremor that generates vibrations on the ocean water's surface can be caused by seismic movement on the seafloor. Most of the time the tremor is caused by the upward or downward movement of a block of oceanic crust that moves a mass of ocean water. A volcanic eruption, meteorite impact, or nuclear explosion can also cause a tsunami.

90%
Movement of tectonic plates

10%
Other causes

RISING PLATE

Water level rises   Water level drops

SINKING PLATE

The displaced water tends to level out, generating the force that causes waves.

7.5

Only earthquakes above this magnitude on the Richter scale can produce a tsunami strong enough to cause damage.

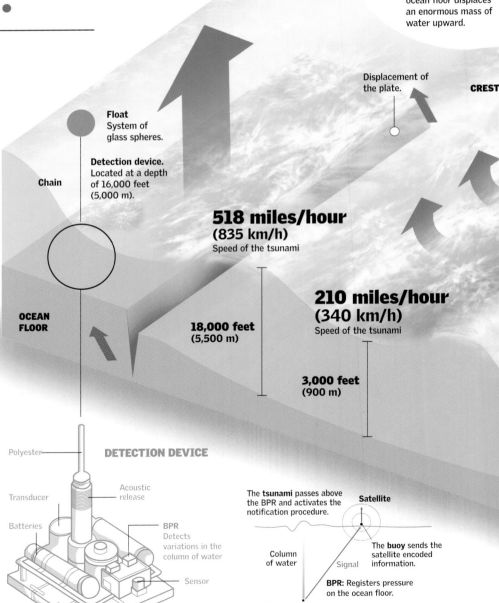

1

THE EARTHQUAKE
A movement of the ocean floor displaces an enormous mass of water upward.

Displacement of the plate.

CREST

Float
System of glass spheres.

Chain

Detection device.
Located at a depth of 16,000 feet (5,000 m).

518 miles/hour
(835 km/h)
Speed of the tsunami

210 miles/hour
(340 km/h)
Speed of the tsunami

OCEAN FLOOR

18,000 feet
(5,500 m)

3,000 feet
(900 m)

Polyester

DETECTION DEVICE

Transducer

Acoustic release

Batteries

BPR
Detects variations in the column of water

Sensor

The **tsunami** passes above the BPR and activates the notification procedure.

Satellite

The **buoy** sends the satellite encoded information.

Column of water

Signal

BPR: Registers pressure on the ocean floor.

WHEN THE WAVE HITS THE COAST

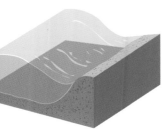

A **Sea level drops abnormally low.**
Water is "sucked" away from the coast by
the growing wave.

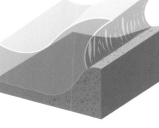

B **The giant wave forms.**
At its highest, the wave
may become nearly
vertical.

COMPARISON OF THE SIZE OF THE WAVE

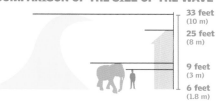

33 feet
(10 m)

25 feet
(8 m)

9 feet
(3 m)

6 feet
(1.8 m)

33 feet
(10 m)

Typical height
a major tsunami
can reach.

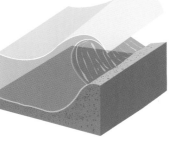

C **The wave breaks along the coast.**
The force of the wave is released
in the impact against the coast.
There may be one wave or several
waves.

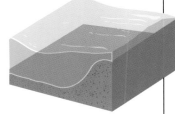

D **The land is flooded.**
The water may take several
hours or even days to return
to its normal level.

② THE WAVES ARE FORMED

As this mass of water drops, the water
begins to vibrate. The waves, however,
are barely 1.5 feet (0.5 m) high, and a
boat may cross over them without the
crew even noticing.

TROUGH

CREST

LENGTH OF THE WAVE
From 62 to 430 miles
(100 to 700 km) on the
open sea, measured from
crest to crest.

③ THE WAVES ADVANCE

Waves may travel thousands of
miles without weakening. As the
sea becomes shallower near the
coast, the waves become closer
together, but they grow higher.

④ TSUNAMI

On reaching the coast,
the waves find their path
blocked. The coast, like a
ramp, diverts all the force
of the waves upward.

Buildings on
the coast may
be damaged or
destroyed.

31 miles per hour
(50 km/h)
Speed of the Tsunami

65 feet
(20 m)

Between 5 and 30
minutes before the
tsunami arrives, the sea
level suddenly drops.

After a catastrophe

The illustration shows a satellite image of Khao Lak, in the coastal province of Phang Nga in southwestern Thailand. The picture was taken three days after the great tsunami of Dec. 26, 2004, that was caused by the colliding Eurasian and Australian plates near Indonesia. This was the greatest undersea earthquake in 40 years. Below, the area as it looked two years earlier.

Everything swept away by the wave was left piled up on the beach.

Coastline before the tsunami

The coastline was devastated.

CAPE PANKARANG

Bamboo Orchid Resort

Palm Galleria Resort

Pankarang Beach Cottage

South Sea Pankarang

Blue Village Pankarang Resort

The tropical forests along the beach of Cape Pankarang were wiped out.

River levels rose after the catastrophe.

The river's mouth was completely flooded.

THAILAND

Surface area	198,114 square miles (513,115 sq km)
Population	63,100,000
Population density	289 people per square mile (1129 people/sq km)
Tsunami-related deaths	5,248
Persons missing in tsunami	4,499

Latitude 16° N
Longitude 100° E

Dec. 29, 2004

The hypothetical line shows how far inland the tsunami reached. Over half a mile (1 km) of the coast was swept away by the tidal wave. Eighty percent of the surface in Khao Lak's tourist area was destroyed.

1,000

Deaths caused by the tsunami in this area alone.

Thailand had the greatest number of tourist deaths.

ANDAMAN SEA

China

Thailand

EPICENTER
OF THE QUAKE

India

Indian Ocean

Khuk Khak Beach

Area flooded
by the wave

3,300-4,900 feet (1,000-1,500 m)

Many beaches
lost all their
sand.

Sofitel Magic
Lagoon Resort

Grand
Diamond
Resort
and Spa

Theptaro Lagoon Resort

1,650
(500)

feet (m)

0
(500)

10:00 a.m.

Time when the wave
hit the beach.

Jan. 13, 2003

Nearly two years before the
tragedy. Wide areas of lush
vegetation, fine, white sand
beaches, and many buildings
enhanced the natural attractions
of this tourism center.

INDIAN OCEAN

Surface area	28.3 million square miles (73.4 million sq km)
Percentage of Earth's surface	14%
Percentage of total volume of the oceans	20%
Length of plate boundaries (in focus)	745 miles (1,200 km)
Countries affected in 2004	21

Duration

The tremor lasted between 8 and 10 minutes, one of the longest on record. The waves took six hours to reach Africa, over 5,000 miles (8,000 km) away.

Cause and Effect

n Dec. 26, 2004, an earthquake occurred that measured 9.0 on the Richter scale, the fifth greatest on record. The epicenter was 100 miles (160 km) off the west coast of Sumatra, Indonesia. This quake generated a tsunami that pummeled all the coasts of the Indian Ocean. The islands of Sumatra and Sri Lanka suffered the worst effects. India, Thailand, and the Maldives also suffered damage, and there were victims as far away as Kenya, Tanzania, and Somalia, in Africa.

ARABIAN PLATE

AFRICA

INDIA
Pop. 1.065 billion
18,045 dead

Vishakhapatnam

0.4 inches per year
(1 cm/year)

Bangalore
Cochin   Madras
Pulmoddai
Batticaloa
Colombo
Matara

SOMALIA
Pop. 8,863,338
289 dead

INDIAN PLATE

SRI LANKA
Pop. 19,905,165
35,322 dead

KENYA
Pop. 34,707,817
1 dead

MALDIVES
Pop. 339,330
108 dead

AFRICAN PLATE

2h

TANZANIA
Pop. 37,445,392
13 dead

3h

4h

5h

6h

Indian Ocean

500 miles per hour
(800 km/h)
Speed of the first wave

7:58
Local time when the tsunami was unleashed (00:58 universal time)

230,507
Estimated dead

30 percent were children

THE VICTIMS
On this map, the number of confirmed deaths and the number of missing persons in each country are added together, giving an estimated total death toll. In addition, 1,600,000 persons had to be evacuated.

BANGLADESH
Pop. 141,340,476
2 dead

EURASIAN PLATE

ASIA

Dhaka

Calcutta

Mandalay

MYANMAR
Pop. 42,720,196
600 dead

Rangoon (Yangon)

PHILIPPINE PLATE

Pacific Ocean

GULF OF BENGAL

Bangkok

THAILAND
Pop. 64,865,523
8,212 estimated dead

INDONESIA
Pop. 238,452,952
167,736 dead

Phuket

MALAYSIA
Pop. 23,522,482
74 dead

PACIFIC PLATE

Banda Aceh

4 inches per year (10 cm/year)

0.4 inches per year (1.0 cm/yr)

S U M A T R A

0.4 inches per year (1.0 cm/yr)

1h

EPICENTER
3° 18' N
95° 47' E

Magnitude 9
Multiple aftershocks of up to magnitude 7.3.

20 sec.

Sumatra

Banda Aceh

Epicenter

1 **Undersea earthquake**
Displacement of 50 feet (15 m) along the edge of the Indian Plate, 18 miles (30 km) below the seabed.

8 min.

2 **The wave begins**
Large waves are detected northwest and southeast of the epicenter.

24 min.

The tsunami's advance.

A seismic station in Australia detected the seismic movement that later caused the great tsunami that struck the nearest coastlines with waves more than 33 feet (10 m) high. An hour and a half later, the tsunami reached Sri Lanka and Thailand. The tsunami had seven crests, which reached the coasts at 20-minute intervals. By the time the tsunami arrived at the coast of Africa hours later, the waves had been greatly diminished.

The wave reaches land

3 **First impact**
A 33-foot-high (10 m) wave destroys Banda Aceh, Indonesia, reaching 2.5 miles (4 km) inland.

Study and Prevention

Predicting earthquakes is very difficult because of a large number of variables, because no two fault systems are alike. That is why populations that have settled in areas with high seismic risk have developed a number of strategies to help everyone know how to act should the earth begin to shake. California and Japan are examples of densely populated

regions whose buildings, now designed according to a stable construction model, have saved many lives. There children are trained periodically at their schools: they do practice drills, and they know where to look for cover. Experts have learned many things about earthquakes in their attempt to understand the causes of these tremors, but they still are not able to predict when an earthquake will take place. ●

Risk Areas

A seismic area is found wherever there is an active fault, and these faults are very numerous throughout the world. These fractures are especially common near mountain ranges and mid-ocean ridges. Unfortunately, many population centers were built up in regions near these dangerous places, and, when an earthquake occurs, they become disaster areas. Where the tectonic plates collide, the risk is even greater. ●

Arctic Ocean

ASIA

6.8
Kobe, 1995
The city of Kobe and nearby villages were destroyed in only 30 seconds.

8.7
Assam, 1897
More than 1,600 people died in northwest India.

Indo-Australian Plate

Pacific Ocean

Mountain

Trench

Pacific Plate

Subduction zone

9.2
Alaska, 1964
Lasted between three and five minutes and caused a tsunami responsible for 122 deaths.

Rocky Mountains

8.3
San Francisco, 1906
Major fires contributed to the devastation of the city.

INDO-AUSTRALIAN PLATE

PACIFIC PLATE

9.0
Sumatra, 2004
Tsunami in Asia
An earthquake near the island of Sumatra created 33-foot (10-m) waves and a human tragedy.

Himalayas

MARIANA TRENCH
The deepest marine trench on the planet, with a depth of 35,872 feet (10,934 m) below sea level. It is on the western side of the north Pacific and east of the Mariana Islands.

PACIFIC PLATE

8.1
Mexico, 1985
Two days later there was a 7.6 aftershock. More than 11,000 people died.

Pacific Ocean

COCOS AND CARIBBEAN PLATES
Contact between these two plates is of the convergent type: the Cocos Plate moves under the Caribbean Plate, a phenomenon known as subduction. This causes a great number of tremors and volcanoes.

Cocos Plate

Caribbean Plate

Indian Ocean

FIJI PLATE

Fiji Plate

AUSTRALIAN PLATE

NEW ZEALAND FAULT
A large fault that moves horizontally, crosses the lithosphere, accommodating the movement between two large crust plates; it is a special type of directional plate called a transforming plate.

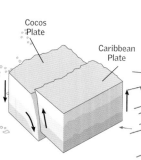

Pacific Plate

ANTARCTIC PLATE

Most-vulnerable regions

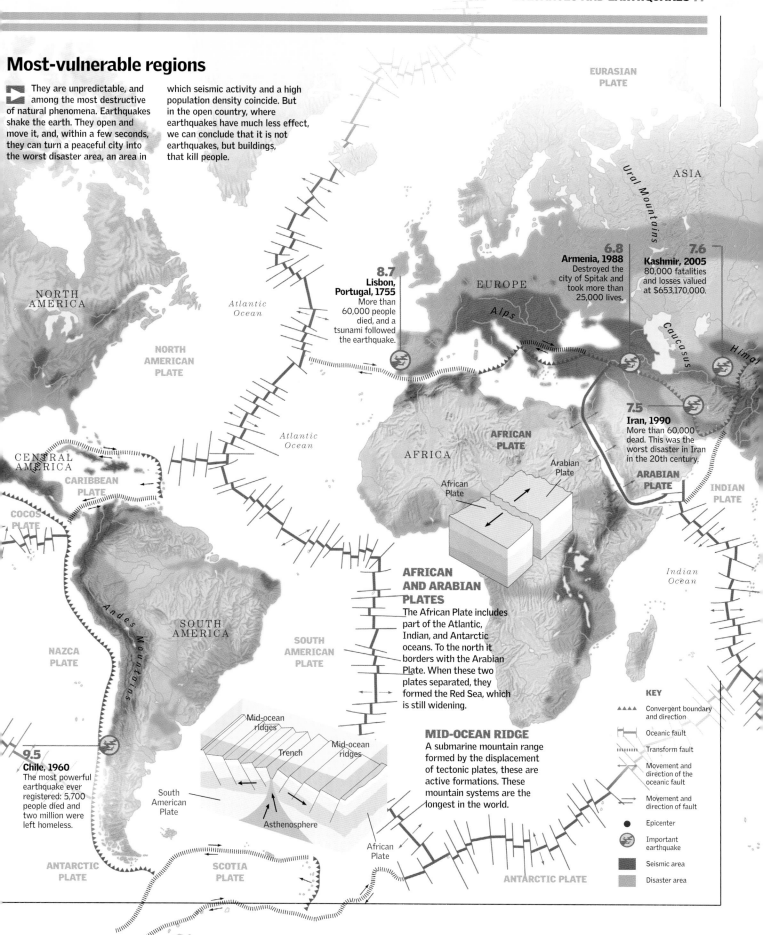

They are unpredictable, and among the most destructive of natural phenomena. Earthquakes shake the earth. They open and move it, and, within a few seconds, they can turn a peaceful city into the worst disaster area, an area in which seismic activity and a high population density coincide. But in the open country, where earthquakes have much less effect, we can conclude that it is not earthquakes, but buildings, that kill people.

EURASIAN PLATE

Ural Mountains

ASIA

NORTH AMERICA

Atlantic Ocean

NORTH AMERICAN PLATE

8.7
Lisbon, Portugal, 1755
More than 60,000 people died, and a tsunami followed the earthquake.

EUROPE

Alps

6.8
Armenia, 1988
Destroyed the city of Spitak and took more than 25,000 lives.

7.6
Kashmir, 2005
80,000 fatalities and losses valued at $653,170,000.

Caucasus

Himal

Atlantic Ocean

AFRICAN PLATE

AFRICA

Arabian Plate

African Plate

7.5
Iran, 1990
More than 60,000 dead. This was the worst disaster in Iran in the 20th century.

ARABIAN PLATE

INDIAN PLATE

CENTRAL AMERICA

CARIBBEAN PLATE

COCOS PLATE

AFRICAN AND ARABIAN PLATES

The African Plate includes part of the Atlantic, Indian, and Antarctic oceans. To the north it borders with the Arabian Plate. When these two plates separated, they formed the Red Sea, which is still widening.

Indian Ocean

Andes Mountains

SOUTH AMERICA

SOUTH AMERICAN PLATE

NAZCA PLATE

Mid-ocean ridges

Trench

Mid-ocean ridges

MID-OCEAN RIDGE

A submarine mountain range formed by the displacement of tectonic plates, these are active formations. These mountain systems are the longest in the world.

9.5
Chile, 1960
The most powerful earthquake ever registered: 5,700 people died and two million were left homeless.

South American Plate

Asthenosphere

African Plate

KEY

▲▲▲▲	Convergent boundary and direction
⊢⊢	Oceanic fault
⊪⊪⊪	Transform fault
⟶	Movement and direction of the oceanic fault
⟶	Movement and direction of fault
●	Epicenter
🌀	Important earthquake
▇	Seismic area
▒	Disaster area

ANTARCTIC PLATE

SCOTIA PLATE

ANTARCTIC PLATE

Precision Instruments

The destructive potential of earthquakes gave rise to the need to study, measure, and record them. Earthquake records, called "seismograms," are produced by instruments called "seismographs," which basically capture the oscillations of a mass and transform them into signals that can be measured and recorded. An earthquake is usually analyzed by means of three seismographs, each oriented in a unique direction at a given location. In this way one seismograph detects the vibrations produced from north to south, another records those from east to west, and a third detects vertical vibrations, those that go up and down. With these three instruments, a seismic event can be reconstructed. ●

Seismometers in History

Modern seismometers have digital mechanisms that provide maximum precision. The sensors are still based on seismic energy moving a mechanical part, however, and that is essentially the same principle that operated the first instrument used to evaluate earthquakes. It was invented by a Chinese mathematician almost 2,000 years ago. Beginning with his invention, the mechanism has been perfected to what it is today.

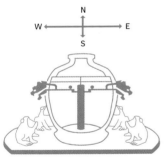

HOW IT WORKS The oscillating mass vibrates when an earthquake takes place. The "dragons," joined to the pendulum by a rigid bar, hold small balls in a delicate equilibrium.

Seismic Wave

1950

PORTABLE SEISMOMETER
Their strong structure allowed these seismometers to be installed in the field. This model translated movement to electric impulses so the signal could be transmitted over some distance.

123

HENG'S SEISMOSCOPE
The first known seismometer was Chinese. The metallic pendulum mass hung from the cover of a large bronze jar. The small balls fell from the mouths of the dragons to the mouths of the frogs, depending on the direction of the seismic movement. Some of these models were 6 feet (2 m) tall.

ZHANG HENG
Chinese mathematician, astronomer, and geographer (AD 78-139), also invented the odometer, calculated the number pi as the square root of 10 (3.16), and corrected the calendar.

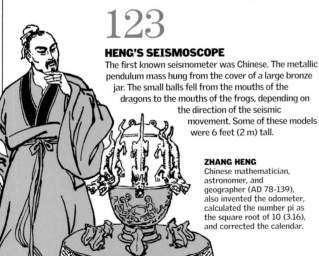

1906

BOSCH-OMORI SEISMOMETER
Is a horizontal pendulum with a pen that makes a mark directly on a paper roll. With it, Omori, a Japanese scientist, registered the 1906 earthquake in San Francisco.

1980

WILMORE PORTABLE SEISMOMETER

A sensitive mass vibrates and moves to the rhythm of the seismic energy inside this tube-shaped mechanism. An electromagnet translates this vibration to electric signals, which are transmitted to a computer that records the data.

Pioneers of seismology

The defining principle of modern seismology emerged from relating earthquakes to the movement of the continents, but that did not take place until well into the 20th century. Starting in the 19th century, however, many scholars contributed elements that would be indispensable.

ROBERT MALLET
From Dublin, Ireland (1810-81). Carried out important studies on the speed of the propagation of earthquakes, even before having experienced one.

JOHN MILNE
British geologist and engineer (1850-1913), created a needle seismograph, a forerunner of current seismographs, and related earthquakes to volcanism.

RICHARD OLDHAM
British (1858-1936). Published a study in 1906 on the transmission of seismic waves, in which he also proposed the existence of the Earth's core.

SPRING
Allows the pivot and mass to move vertically (vibrating movement).

SEISMOGRAM
The record of amplitude on the paper strip.

OSCILLATING PENCIL
Moves to the rhythm of the vibrations amplified by the mechanism..

ROTATING DRUMS
Move the roll at a constant and precise speed.

CLOCK AND RECORDER
Take the signal, synchronize it, and convert it.

PIVOT
Supports and maintains an axis. It can have a hinge.

SUSPENDED MASS
Moves according to the direction of the waves of the earthquake and in proportion to their strength.

HORIZONTAL MOVEMENT

MOVEMENT SENSOR
The floating mass is displaced and moves a part inside an electromagnet. Variations produced in the magnetic field are converted into signals.

HORIZONTAL MOVEMENT

How a Seismograph Works

The Earth's tremors produce movements in the mass that serves as a sensor. If the pivot is hinged, it allows movements in only one direction: horizontally or vertically, depending on the sensitivity and calibration of the spring. These movements are transformed into electric or digital signals to give versatility in processing and recording the data.

SUSPENSION
The small vibrations of the ground will move the base more.

CONNECTING CABLE
Transmits the electric signals that are generated.

ANCHORED BASE
The greater the degree of suspension, the greater the sensitivity of the mechanism.

Continuous Movement

Humankind has tried throughout history to find a way to predict earthquakes. Today, this is done through the installation of seismic observatories and of various field instruments that gather information and compare it to the data sent by scientists from other locations. Based on these records, it is possible to evaluate the chances that a great earthquake is developing and act accordingly. ●

Observing from a Distance

Seismologists place instruments at fault lines in earthquake-prone areas. Later, at the seismologic observatory, the information taken by field instruments is compiled, and any significant change is noted. If anything suggests that an earthquake is about to take place, emergency services are alerted. Most of these instruments are automatic, and they send digital data through the telephone system.

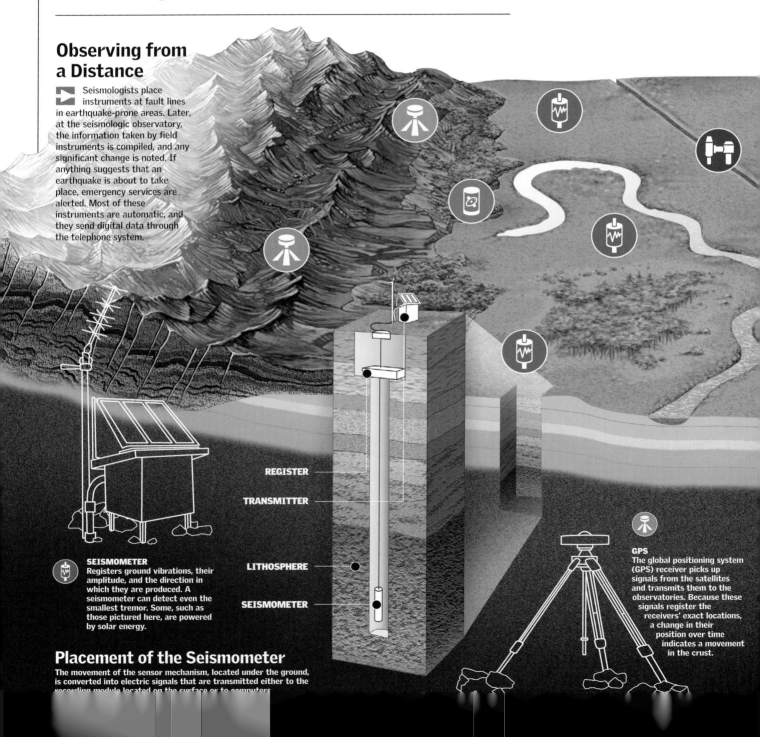

REGISTER

TRANSMITTER

LITHOSPHERE

SEISMOMETER

SEISMOMETER
Registers ground vibrations, their amplitude, and the direction in which they are produced. A seismometer can detect even the smallest tremor. Some, such as those pictured here, are powered by solar energy.

GPS
The global positioning system (GPS) receiver picks up signals from the satellites and transmits them to the observatories. Because these signals register the receivers' exact locations, a change in their position over time indicates a movement in the crust.

Placement of the Seismometer
The movement of the sensor mechanism, located under the ground, is converted into electric signals that are transmitted either to the recording module located on the surface or to computers.

Seismologic Networks

Installing complex detection systems would not be of much use if the systems worked in an isolated manner and were not able to share the information they generated. There are national and international seismologic networks that, by means of communications technology, send their observations to other areas that might be affected.

NETWORK OF NETWORKS
Findings in an area can have repercussions at a great distance. The immediate availability of data allows for linked work.

SATELLITES
Some are used by the GPS systems, but others are critical because they take photographs with extreme accuracy, and they are thus able to record indicators that can be communicated quickly to the base.

LABORATORY
Networking at the research centers allows for the comparison of data and provides a global vision that broadens the predictive power of science.

CREEP METER
Measures the movement between earthquakes or time interval between the two boundaries of a fault. It includes a tension system with a calibrated mechanism. Any movement between the ends alters the magnetic field.

MAGNETOMETER
The magnetic field of the Earth changes when the tensions between rocks vary. Therefore, a change in magnetism can indicate tectonic movement. The magnetometer can distinguish between these changes and those that are more general.

LEVELED GROUND

LEVELED GROUND

FAULT

CABLE HOUSING

Earthquakes cannot be predicted

For a prediction system to be acceptable, it must be accurate and trustworthy. Therefore, it must have a small level of uncertainty regarding location and the timing, and it must minimize errors and false alarms. The cost of evacuating thousands of people, of providing lodging for them, and of making up for their loss of time and work for a false alarm would be unrecoverable. At this time there is no trustworthy method for predicting earthquakes.

Placement of a Creep Meter

To measure the relative movement between the ends, two posts are fixed, one at each side of the fault, 6 feet (2 m) under the ground, or over the concrete base, at a fixed angle (but not at a right angle).

Stable Buildings

In cities located in seismic areas, buildings must be designed and constructed with an earthquake-resistant structure that can adequately withstand the movements caused by an earthquake. Foundations are built with damping so that they can absorb the force of the seismic movement. Other buildings have a large metallic axis, around which the stories of the building can oscillate without falling. Currently the amount of knowledge on the effect of earthquakes on structures, as well as knowledge on the behavior of different materials, allows for the construction of less-vulnerable buildings. ●

Why Pagodas in Japan Do Not Fall

Japanese pagodas have survived centuries of earthquakes. They are five stories tall, and higher sections of the building are smaller than lower parts. The pagodas are held up by a central pillar that acts as the only support for the building. During a quake, each floor balances independently, without transmitting the oscillating force to the other floors.

Swaying Principle
Each floor also has a symmetric structure. During an earthquake, the ends opposite the eaves act as a counterbalance to achieve equilibrium.

ROPPONGI HILLS TOWER
Located in Tokyo, its structure consists of simple geometric bodies, symmetrically placed, without any irregularities in shape. The tower is formed by a massive central framework and a lightweight and flexible exterior framework.

898 FEET (274 M) 56 STORIES

Central pillar

ROPPONGI HILLS
TOKYO, JAPAN

Latitude 36° N
Longitude 140° E

Covered area	4,089,930 square feet (380,105 sq m)
Stories	54 (+6 underground)
Width	275 feet (84 m)
Height	780 feet (238 m)

GOJU Fifth floor
SHIJU Fourth floor
SANJU Third floor
NIJU Second floor
SHOJU First floor
SHINBASIRA Central pillar

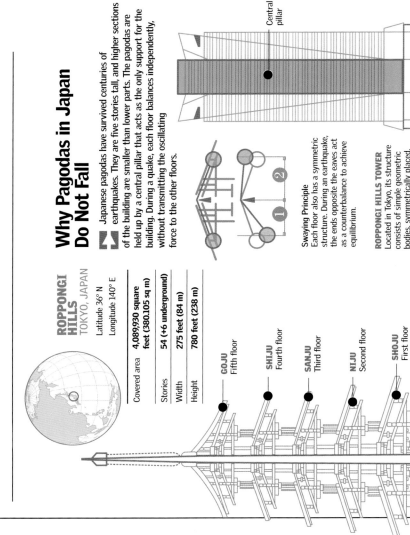

Earthquake-Resistant Architecture

There are many ways to design an earthquake-resistant structure: the distribution of walls, the joints between beams and columns, and geometric simplicity. There are also earthquake simulators, large platforms that shake a structure to test it. The simulators are used to test materials and study the forces that act on them. A building's true earthquake resistance, though, can be proven only when it has been built and has survived actual earthquakes.

STRUCTURE
To avoid imbalances, the upper elements of a structure must be located over only a few axes, without any isolated vertical segments.

AVOIDANCE OF OFF-CENTER JOINTS
If the beam remains still when the wall moves, the joint breaks. Forces spread out over an axis with flexible material.

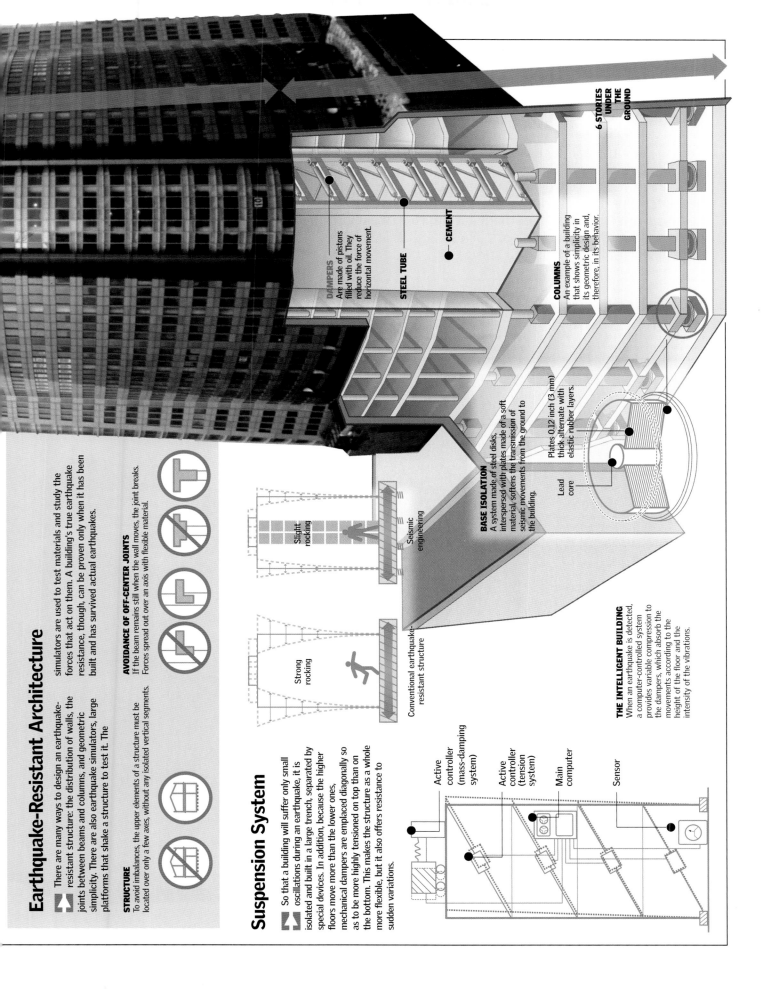

DAMPERS
Are made of pistons filled with oil. They reduce the force of horizontal movement.

STEEL TUBE

CEMENT

6 STORIES UNDER THE GROUND

COLUMNS
An example of a building that shows simplicity in its geometric design and, therefore, in its behavior.

BASE ISOLATION
A system made of steel disks, interspersed with plates made of a soft material, softens the transmission of seismic movements from the ground to the building.

Plates 0.12 inch (3 mm) thick alternate with elastic rubber layers.

Lead core

Suspension System

So that a building will suffer only small oscillations during an earthquake, it is isolated and built in a large trench, separated by special devices. In addition, because the higher floors move more than the lower ones, mechanical dampers are emplaced diagonally so as to be more highly tensioned on top than on the bottom. This makes the structure as a whole more flexible, but it also offers resistance to sudden variations.

Slight rocking

Seismic engineering

Strong rocking

Conventional earthquake-resistant structure

THE INTELLIGENT BUILDING
When an earthquake is detected, a computer-controlled system provides variable compression to the dampers, which absorb the movements according to the height of the floor and the intensity of the vibrations.

Active controller (mass-damping system)

Active controller (tension system)

Main computer

Sensor

On Guard

When the earth shakes, nothing can stop it. Disaster seems inevitable, but, though it is inevitable, much can be done to diminish the extent of the catastrophe. Residents of earthquake-prone areas have incorporated a series of preventive measures to avoid being surprised and to help them act appropriately at home, at the office, or outdoors. These are basic rules of behavior that will help you survive. ●

Prevention

If you live in an earthquake-prone area, familiarize yourself with the emergency plans for the community where you live, plan how your family will behave in the event of an earthquake, know first aid, and know how to extinguish fires.

FIRST-AID KIT
Keep a first-aid kit, and keep your vaccinations up to date.

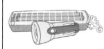

LIGHTS
Have emergency lighting, flashlights, a transistor radio, and batteries on hand.

SECURING OBJECTS
Secure heavy objects such as furniture, shelves, and gas appliances to the wall or to the floor.

BREAKERS
Have a breaker installed, and know how to shut off the electricity and the gas supply.

FOOD AND WATER
Store drinking water and nonperishable food.

FIRST AID
Learn first aid, and participate in community earthquake-response training.

During an earthquake

As soon as you feel the earth under your feet begin to move, look for a safe place, such as beneath a doorframe or under a table, to take cover. If you happen to be on a street, head to an open space such as a square or park. It is important to remain calm and to not be influenced by people who panic.

AT HOME
It is essential that the home be built following regulations for earthquake-resistant construction and that someone be in charge of shutting off the electricity and gas supplies.

Objects that could fall because of movement should be attached to the wall.

Do not use elevators: the electricity might be cut off.

In case of evacuation, stairs are the safest place, but they could become filled with people.

It is good to designate a leader who can guide others. Form a human chain to prevent getting lost and to prevent accidents.

Toxic or flammable materials must not be in danger of spilling.

Determine safe spots under doorframes, next to a pillar, or under a table.

Do not light matches or flames: use flashlights.

Do not drink tap water because it might be polluted.

Fires are frequently more dangerous than the earthquake itself. They can easily get out of control and spread through the city.

Do not run on the street; doing so will cause panic.

AT THE OFFICE

Offices are usually located in areas conducive to bringing large groups of people together. Thus it is recommended that you remain where you are and not rush to the exits. When people panic, there is a greater probability of their being crushed by a crowd than by a building, especially in buildings that contain a lot of people.

IN PUBLIC PLACES

When you are outside, it is important to keep away from tall buildings, light poles, and other objects that could fall. The safest course of action is to head to a park or other open space. If the earthquake takes you by surprise while you are driving, stop and remain in the car, but make sure you are not close to any bridges.

Mark escape routes and keep them free of obstacles.

Know where emergency equipment, such as fire extinguishers, hoses, and axes, is located.

Head toward open spaces such as squares and parks, and move away from any trees to the extent possible.

Coastlines

Do not approach the coastline because of the possibility of a tsunami. Also avoid rivers, which could develop strong currents.

Seek protection under a table or desk to avoid being hurt by falling objects.

Stay away from windows and balconies.

Follow the instructions of civil defense officers.

Fixing breaks in water and gas lines is a priority.

Stay away from buildings, walls, utility poles, and other objects that could fall.

If you are near an exit, leave the building and walk away. Do not block doorways.

Rescue Tasks

Once the earthquake ends, rescue tasks must begin. At this stage it is imperative to determine whether anyone is injured and to apply first aid. Do not move injured people who have fractures, and do not drink water from open containers.

If you are in a vehicle, stop in the safest place possible (away from large buildings, bridges, and utility poles). Do not leave your car unless it becomes necessary to do so.

RESCUERS
The first priority after an earthquake is to search for survivors..

DOGS
Specially trained animals with protective helmets and masks can search for people under the rubble.

TRANSPORTATION
It is important to keep access routes to affected areas open to ensure entry by emergency teams.

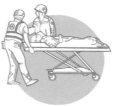

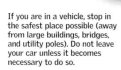

San Francisco in Flames

The earthquake that shook San Francisco on April 18, 1906, was a major event: in only a few seconds, a large part of one of the most vital cities of the United States was reduced to rubble. Suddenly, centuries of pent-up energy was released when the earthquake, measuring 8.3 on the Richter scale, devastated the city. Though the earthquake destroyed many buildings, the worst damage was caused by the fires that destroyed the city in the course of three days, forcing people to flee their homes.

April 18

1 **EVERYTHING STARTED LIKE THIS.**
On April 18, 1906, at 5:12 a.m., the Pacific Plate experienced movement of approximately 19 feet (6 m) along its 267-mile (430-km) length along the northern San Andreas fault. The earthquake's epicenter lay 39 miles (64 km) north of San Francisco. In seconds, the earth began to move, and the majority of the city's buildings collapsed. The trolleys and carriages that were moving through the cobblestoned streets of the city were reduced to rubble.

CITY HALL
The facade was crowned by a dome that was supported by a system of columns on a steel structure. It was considered one of the city's most beautiful buildings.

GAS LIGHTING
Gas lighting was one of the signs of progress that gave prominence to the city.

History of City Hall

Until the earthquake struck, City Hall had been the seat of city government and the symbol of the city. Built in the second half of the 19th century, it represented a time of accelerated growth, powered by the gold riches of the state of California. Construction began on Feb. 22, 1870, and ended 27 years later, after many revisions to architect Auguste Laver's original project. While it stood, City Hall was said to have been constructed with bricks held together with corruption, typical of a time of easy money and weak institutions. The total cost of the work rose to a little more than $6,000,000 of that time, and, according to current calculations, it is estimated that it was prepared to withstand an earthquake up to a magnitude of 6.6. Only the dome and the metal structure were left standing. The remnants of the building were demolished in 1909.

THE FACE OF THE BUILDING COLLAPSED.
The facade collapsed completely on top of the rotunda at its base.

UNITED STATES SAN FRANCISCO, CALIFORNIA

Latitude 42° 40' N
Longitude 122° 18' W

Surface area	**46 square miles (120 sq km)**
Population	**739,426**
Population density	**16,000 people per square mile (6,200 people/sq km)**
Perceptible earthquakes felt annually	**100 - 150**
Total earthquakes per year	**˜ 10,000**

April 20

3 **THE GREAT FIRE**
Two days later, what had begun as a localized fire had become an inferno that consumed the city. There were mass evacuations of people to distant areas, while the army dynamited some buildings. Firefighters had to control the flames using seawater.

3 years

4 **REBUILDING**
The city reemerged from the ashes, powered by its wealth and economic importance. Losses are estimated to have reached $5,000,000,000 in present-day dollars, and, until Hurricane Katrina struck in 2005, the 1906 earthquake was the greatest natural disaster the United States had experienced.

5:12 a.m.

2 **THE EARTHQUAKE**
Not only was the earthquake extremely violent, but it oscillated in every direction for 40 seconds. People left their houses and ran down the streets, completely stunned and blinded by fear. Many buildings split open, and others became piles of rubble. A post office employee related that "The walls were thrown into the middle of the rooms, destroying the furniture and covering everything with dust."

Clearing the Rubble

It is calculated that some 3,000 people died in the 1906 catastrophe, trapped in their destroyed homes or burned by the fire the earthquake started. In the following weeks, the army, firefighters, and other workers deposited the rubble in the bay, forming new land, which is today known as the Marina District. Little by little, traffic resumed in the major streets, and the trolley system was reestablished. Six weeks after the earthquake, banks and stores opened for business.

FIELD LUNCH
The army set up kitchens in the camps. There was always food in these field kitchens, and there was even a free ration of tobacco for every person.

Workers pose while they demolish a house.

Army tents to house refugees.

WORKERS
By Saturday, April 21, some 300 plumbers had entered the city to reestablish services, mainly the water system. During the following weeks, thousands of workers tore down unstable buildings, prepared the streets for traffic, and cleared the city of rubble. Nearly 15,000 horses were used to haul rubble.

SHORTLY AFTERWARD
This panoramic photograph shows the destruction of the city. Despite the destruction, many buildings were left standing.

SAINT MARY'S CHURCH

CHINATOWN
Was completely destroyed by the fire.

18,000

BUILDINGS OF THAT PERIOD
are still standing, despite the 1906 earthquake and the tremors that passed through the city afterward.

Three Days of Fire

The great fire that followed the earthquake expanded quickly. Firefighters, in a desperate attempt to block the spread of the fire, used explosives to make firebreaks because there was no water supply available. The army evacuated the area, and people could not take anything with them. During the three days when the city burned, it is speculated that many homeowners burned their houses that had been partially destroyed by the earthquake, in order to be able to collect insurance money. Other things that contributed to the fire were the intentional explosions that, at first badly implemented, spread the fire. By the fourth day, the center of the city was reduced to ashes.

1 The fire began in the Market Street area, south of the city in the worker's district, where many houses were made of wood.

2 On the second day, the fire spread west. About 300,000 people were evacuated from the bay in ferries.

Day 3
Day 2
Day 1
Day 2

3 During the third day, the fire swept through Chinatown and North Beach, causing heavy damage to the Victorian homes on North Beach hill.

4 Once the fire was extinguished, Russian Hill and Telegraph Hill (shown as white spots) were still intact, as was the port.

REFUGEE CAMPS

The army set up field camps in the parks to house those who had lost everything. Months later, the government built temporary homes for about 20,000 people.

The firefighters tried to extinguish the fire.

MERCHANT EXCHANGE
Built in 1903, it remained standing and was later refurbished.

MILLS BUILDING
This building in the financial district had been built in 1890.

28,000

BUILDINGS DEMOLISHED
The damage calculated to have been caused by the great earthquake. Many of these buildings, such as City Hall, were famous for their lavish architecture.

Historic Earthquakes

The Earth is alive. It moves, it shifts, it crashes and quakes, and it has done so since its origin. Earthquakes vary from a soft vibration to violent and terrorizing movements. Many earthquakes have gone down in history as the worst natural catastrophes ever survived by humanity. Lisbon, Portugal, 1755; Valdivia, Chile, 1960; and Kashmir, Pakistan, 2005, are only three examples of the physical, material, and emotional devastation in which an earthquake can leave a population. ●

1995

KOBE, JAPAN

Magnitude	6.8 (Richter)
Fatalities	6,433
Material losses	$100 billion

AN INFERNO
The great earthquake of Hanshin that occurred on Jan. 17, 1995, in Kobe, a Japanese port, left behind more than 6,000 dead, 38,000 injured, and 319,000 people who had to be housed in more than 1,200 emergency shelters. The Nagata District was one of the hardest-hit areas. Almost 80 percent of the victims died because the old wooden homes crumbled in the generalized fires that followed the earthquake.

AFTER THE HORROR. The world was shaken, looking at the horrendous images of how Kobe, the city by the sea, had been painfully broken to pieces.

1755

LISBON, PORTUGAL

Magnitude	8.7 (Richter)
Fatalities	62,000
Material losses	unknown

It was the Day of the Dead, and, at 9:20 in the morning, almost the entire population of Lisbon was at church. While mass was celebrated, the earth quaked, and this earthquake would be one of the most destructive and lethal in history. The earthquake unleashed a tsunami that was felt from Norway to North America and that took the lives of those who had sought shelter in the river.

1906

SAN FRANCISCO, U.S.

Magnitude	8.3 (Richter)
Fatalities	3,000
Material losses	$5 billion

The city was swept by the earthquake and by the fires that followed it. The quake was the result of the rupture of more than 40 miles of the San Andreas fault. It is the greatest earthquake in the history of the United States: 300,000 people were left homeless, and property losses reached millions in 1906 dollars. Buildings collapsed, the fires spread for three days, and the water lines were destroyed.

2004

SUMATRA, INDONESIA

Magnitude	9.0 (Richter)
Fatalities	230,507
Material losses	incalculable

An earthquake whose epicenter crossed the island of Sumatra, Indonesia, took place on December 26. This earthquake generated a tsunami that affected the entire Indian Ocean, primarily the islands of Sumatra and Sri Lanka, and reached the coasts of India, Thailand, the Maldives, and even Kenya and Somalia. This was a true human tragedy, and the economic damages were incalculable.

1960

VALDIVIA, CHILE

Magnitude	9.5 (Richter)
Fatalities	5,700
Material losses	$500 million

Known as the Great Chilean Earthquake, this was the strongest earthquake of the 20th century. The surface waves produced were so strong that they were still being registered by seismometers 60 hours after the earthquake. The earthquake was felt in various parts of the planet, and a huge tsunami spread through the Pacific Ocean, killing more than 60 people in Hawaii. One of the most powerful earthquakes in memory, its aftershocks lasted for more than a week. More than 5,000 people died, and nearly two million people suffered damage and loss.

1985

MEXICO CITY, MEXICO

Magnitude	8.1 (Richter)
Fatalities	11,000
Material losses	$1 billion

The city shook on September 19. Two days later, there was an aftershock measuring 7.6 on the Richter scale. In addition to 11,000 deaths, there were 30,000 injured, and 95,000 people were left homeless. As the Cocos Plate slid under the North American Plate, the North American Plate fractured, or split, 12 miles (20 km) inside the mantle. The vibrations of the ocean floor off the southwestern coast of Mexico provoked a tsunami and produced energy 1,000 times as great as that of an atomic bomb. Strong seismic waves reached as far east as Mexico City, a distance of 220 miles (350 km).

2005

KASHMIR, PAKISTAN

Magnitude	7.6 (Richter)
Fatalities	80,000
Material losses	$595 million

Also known as the Indian Subcontinent Earthquake, the North Pakistan Earthquake, and the South Asian Earthquake. It occurred on Oct. 8, 2005, in the Kashmir region between India and Pakistan. Because schools were in session when the earthquake struck (9:20 a.m.), many of the victims were children, who died when their school buildings collapsed. It was the strongest earthquake experienced by the region for a century. Three million people lost their homes. The most-heavily affected areas lost all their cattle. Entire fields disappeared under earth and rock. The epicenter was located near Islamabad, in the mountains of Kashmir, in an area governed by Pakistan.

Glossary

Aa

Type of lava flow that presents sharp projections on its surface when it hardens.

Abrasion

Modification of rock surfaces by friction and by the impact of other particles transported by wind, water, and ice.

Active Volcano

Volcano that erupts lava and gas at regular intervals.

Aerosol

Small particles and drops of liquid scattered in the air by volcanic gases.

Aftershock

Small temblor or quake produced as rock settles into place after a major earthquake.

Aseismic

The characteristic of a building designed to withstand oscillations, or of areas with no seismic activity.

Aseismic Region

Tectonically stable region of the Earth, where there are almost no earthquakes. For example, the Arctic region is aseismic.

Ashfall

Phenomenon in which gravity causes ash (or other pyroclastic material) to fall from a smoke column after an eruption. The distribution of the ash is a function of wind direction.

Asthenosphere

Internal layer of the Earth that forms part of the mantle.

Avalanche

Rapid movement of enormous volumes of rock and other materials caused by instability on the flanks of the volcano. The instability can be caused by the intrusion of magma into the structure of the volcano, by a large earthquake, or by the weakening of the volcano's structure by hydrothermal variation, for example.

Ballistic (Fragment)

A lump of rock expelled forcefully by a volcanic eruption and that follows a ballistic or elliptical trajectory.

Baltic

Of or pertaining to the Baltic Sea, or to the territories along it.

Batholith

Massive body of magma that results from an intrusion between preexisting layers.

Caldera

Large, round depression left when a volcano collapses onto its magma chamber.

Convection Currents

Vertical and circular movement of rock material in the mantle but found exclusively in the mantle.

Convergent Boundary

Border between two colliding tectonic plates.

Core

Central part of the Earth, with an outer boundary 1,800 miles (2,900 km) below the Earth's surface. The core is believed to be composed of iron and nickel—with a liquid outer layer and a solid inner core.

Crater

Depression on the peak of a volcano, or produced by the impact of a meteorite.

Crust

Outermost, rigid part of the Earth, made up mostly of basaltic rocks (underneath the oceans) and of rocks with a higher silicate content (in the continents).

Density

Ratio of a body's mass to its volume. Liquid water has a density of 62.4 pounds per cubic foot (1 g/cu cm).

Dike

Tabular igneous intrusion that crosses through layers of surrounding rock.

Dome

Cup-shaped bulge with very steep sides, formed by the accumulation of viscous lava. Usually a dome is formed by andesitic, dacitic, or rhyolitic lava, and the dome can reach a height of many hundreds of feet.

Duration of Earthquake

Time during which the shaking or tremor of an earthquake is perceptible to humans. This period is always less than that registered by a seismograph.

Earthquake

Vibration of the Earth caused by the release of energy.

Eon

The largest unit of time on the geologic scale, of an order of magnitude greater than an era.

Epicenter

Point on the Earth's surface located directly above the focus of an earthquake.

Epicentral Area

Region around the epicenter of an earthquake, usually characterized by being the area where the shaking is most intense and the earthquake damage is greatest.

Epicentral Distance

Distance along the Earth's surface from the point where an earthquake is observed to the epicenter.

Extinct Volcano

Volcano that shows no signs of activity for a long period of time, considered to have a very low probability of erupting.

Fault Displacement

Slow, gradual movement produced along a fault. It is characterized by not generating an earthquake or tremor.

Focus

Internal zone of the Earth, where seismic waves are released, carrying the energy held by rocks under pressure.

Fumarole

Emission of steam and gas, usually at high temperatures, from fractures or cracks in the surface of a volcano or from a region with volcanic activity. Most of the gas emitted is steam, but fumarole emissions can include gases such as CO_2, CO, SO_2, H_2S, CH_4, HCl, among others.

Geothermal Energy

Naturally heated steam used to generate energy.

Geyser

Spring that periodically expels hot water from the ground.

Gondwana

Southern portion of Pangea, which at one time included South America, Africa, Australia, India, and Antarctica.

Hot Spot

Point of concentrated heat in the mantle capable of producing magma that shoots up to the Earth's surface.

Hydrothermal Alteration

Chemical change in rocks and minerals, produced by an aqueous solution that is rich in volatile chemical elements found at high temperature and that rises from a magma body.

Igneous Activity

Geologic activity involving magma and volcanic activity.

Incandescent

A property of metal that has turned red or white because of heat.

Lahar

Mudflows produced on the slopes of volcanoes when unstable layers of ash and debris become saturated with water and flow downhill.

Lapilli

Fragments of rock with a diameter between 0.06 and 1.3 inches (2 and 32 mm) expelled during a volcanic eruption.

Lava

Magma, or molten rock, that reaches the Earth's surface.

Lava Bombs

Masses of lava that a volcano expels, which have a diameter equal to or greater than 1.2 inches (3.2 cm).

Lava Flow

River of lava that flows out of a volcano and runs along the ground.

Liquefaction

Transformation of ground from solid to fluid state through the action of an earthquake.

Lithosphere

Rigid part of the outer layer of the Earth, formed by the crust and the outer layer of the mantle. This is the layer that is destroyed in subduction zones and that grows in mid-ocean ridges.

Magma

Mass of molten rock deep below the surface, which includes dissolved gas and crystals. When magma has lost its gases and reaches the surface, it is called lava. If magma cools within the Earth's crust, it forms plutonic rocks.

Magma Chamber

Section within a volcano where incandescent magma is found.

Mantle

Layer between the Earth's crust and the outer core. Its lower part, the asthenosphere, is partially molten. The more superficial and less-fluid outer part is called the lithosphere.

Mid-Ocean Ridge

An elongated mountain range on the ocean floor, which varies between 300 and 3,000 miles (500 and 5,000 km) in breadth.

Neck

Column of lava that has solidified inside a volcano.

Normal Fault

Fracture in rock layers where the ground is being stretched, which generally causes the upper edge to sink relative to the lower part.

Ocean Trench

Long, narrow, extremely deep area of the ocean floor formed where the edge of an oceanic tectonic plate sinks beneath another plate.

Pahoehoe Lava

Lava with a smooth surface that has a ropelike form.

Pelean Eruption

Type of volcanic eruption with a growing dome of viscous lava that may be destroyed when it collapses because of gravity or brief explosions. Pelean eruptions produce pyroclastic flows or burning clouds. The term comes from Mount Pelée in Martinique.

Permeable Layers

Strata of the Earth's crust that allow water to reach deeper layers.

Plate Tectonics

Theory that the Earth's outer layer consists of separate plates that interact in various ways, causing earthquakes and forming volcanoes, mountains, and the crust itself.

Plinian Eruption

Extremely violent and explosive type of volcanic eruption that continuously expels large quantities of ash and other pyroclastic materials into the atmosphere, forming an eruption column typically 5 to 25 miles (8 to 40 km) high. The term honors Pliny the Younger, who observed the eruption of Mount Vesuvius (Italy) in AD 79.

Plume

Column of hot rock that rises from within the mantle, inside of which the rock may melt.

Primary (P) Wave

Seismic wave that alternately compresses and stretches the ground along its direction of travel.

Pumice

Pale volcanic rock full of holes, which give it a low density. Its composition is usually acidic (rhyolitic). The holes are formed by volcanic gases that expand as volcanic material rises to the surface.

Pyroclastic Flow

Dense, hot mix of volcanic gas, ash, and rock fragments that flows rapidly down the sides of a volcano.

Reverse Fault

Fractures in rock layers where the ground is being compressed, which generally causes the upper edge to rise above the lower part in a plane inclined between 45 and 90 degrees from the horizontal.

Richter Scale

Measures the magnitude of an earthquake or of the energy it releases. The scale is logarithmic, such that an earthquake of magnitude 8 releases 10 times as much energy as a magnitude 7 quake. An earthquake's magnitude is estimated based on measurements taken by seismic instruments.

Rift Zone

Area where the crust is splitting and stretching, as shown by cracks in the rock. Such areas are produced by the separation of tectonic plates, and their presence causes earthquakes and recurrent volcanic activity.

Scale of Intensity

Scale used to measure the severity of movement of the ground produced by an earthquake. Degrees of intensity are assigned subjectively depending on how the tremor is perceived and according to the damage caused to buildings. A widely used scale is the Mercalli scale.

Secondary (S) Wave

Transverse or cross-section wave with motion perpendicular to the direction of its travel.

Seismic Event

Shaking of the ground caused by an abrupt and violent movement of a mass of rock along a fault, or fracture, in the crust. Active volcanoes cause a wide variety of seismic events.

Seismic Gap

Fault zone, or zone of a segment at the boundary between tectonic plates, with a known seismic history and activity, which records a period of prolonged calm, or of seismic inactivity, during which large amounts of elastic energy of deformation accumulate, and that, therefore, presents a greater probability of rupture and occurrence of a seismic event.

Seismic Hazard Calculation

Process of determining the seismic risk of various sites in order to define areas with similar levels of risk.

Seismic Risk

The probability that the economic and social effects of a seismic event will exceed certain preestablished values during a given period, for example, a certain number of victims, an amount of building damage, economic losses, etc. Also defined as the comparative seismic hazard of one site relative to another.

Seismic Wave

Wavelike movement that travels through the Earth as a result of an earthquake or an explosion.

Seismic Zone

Limited geographic area within a seismic region, with similar seismic hazard, seismic risk, and earthquake-resistant design standards.

Seismograph

Instrument that registers seismic waves or tremors in the Earth's surface during an earthquake.

Seismology

Branch of geology that studies tremors in the Earth, be they natural or artificial.

Shield Volcano

Large volcano with gently sloping flanks formed by fluid basaltic lava.

Silicon

One of the most common materials, and a component of many minerals.

Subduction

Process by which the oceanic lithosphere sinks into the mantle along a convergence boundary. The Nazca Plate is undergoing subduction beneath the South American Plate.

Subduction Zone

Long, narrow region where one plate of the crust is slipping beneath another.

Surface Wave

Seismic wave that travels along the Earth's surface. It is perceived after the primary and secondary waves.

Swarm of Earthquakes

Sequence of small earthquakes that occur in the same area within a short time period, with a low magnitude in comparison to other earthquakes.

Symmetry

Correspondence that exists in an object with respect to a center, an axis, or a plane that divides it into parts of equal proportions.

Tectonic Plates

Large, rigid sections of the Earth's outer layer. These plates sit on top of a more ductile and plastic layer of the mantle, the asthenosphere, and they drift slowly at an average rate of 1 inch (2.4 cm) or more per year.

Thrust Fault

A fracture in rock layers that is characterized by one boundary that slips above another at an angle of less than 45 degrees.

Transform Fault

Fault in which plate boundaries cause friction by sliding past each other in opposite directions.

Tremor

Seismic event perceived on the Earth's surface as a vibration or shaking of the ground, without causing damage or destruction.

Tsunami

Word of Japanese origin that denotes a large ocean wave caused by an earthquake.

Viscous

Measure of a material's resistance to flow in response to a force acting on it. The higher the silicon content, the higher the viscosity.

Volcanic Glass

Natural glass formed when molten lava cools rapidly without crystallizing. A solid-like substance made of atoms with no regular structure.

Volcanic Ring

Chain of mountains or islands located near the edges of the tectonic plates and that is formed as a result of magma activity associated with subduction zones.

Volcano

Mountain formed by lava, pyroclastic materials, or both.

Volcanology (Vulcanology)

Branch of geology that studies the form and activity of volcanoes.

Vulcanian Eruption

Type of volcanic eruption characterized by the occurrence of explosive events of brief duration that expel material into the atmosphere to heights of about 49,000 feet (15 km). This type of activity is usually linked to the interaction of groundwater and magma (phreatomagmatic eruption).

Water Spring

Natural source of water that flows out of the crust. The water comes from rainwater that seeps into the ground in one place and comes to the surface in another, usually at a lower elevation. Because the water is not confined in waterproof chambers, it can be heated by contact with igneous rock. This causes it to rise to the surface as hot springs.

Index

W

Y